Wheat

Production, Properties and Quality

Wheat

Production, Properties and Quality

Edited by

W. BUSHUK
Department of Food Science
University of Manitoba
Winnipeg

and

V. F. RASPER
Department of Food Science
University of Guelph
Guelph

Published by
Blackie Academic and Professional, an imprint of Chapman & Hall,
Wester Cleddens Road, Bishopbriggs, Glasgow G64 2NZ

Chapman & Hall, 2–6 Boundary Row, London SE1 8HN, UK

Blackie Academic & Professional, Wester Cleddens Road, Bishopbriggs, Glasgow G64 2NZ, UK

Chapman & Hall GmbH, Pappelallee 3, 69469 Weinheim, Germany

Chapman & Hall USA., One Penn Plaza, 41st Floor, New York NY 10119, USA

Chapman & Hall Japan, ITP-Japan, Kyowa Building, 3F, 2-2-1 Hirakawa-cho, Chiyoda-ku, Tokyo 102, Japan

DA Book (Aust.) Pty Ltd, 648 Whitehorse Road, Mitcham 3132, Victoria, Australia

Chapman & Hall India, R. Seshadri, 32 Second Main Road, CIT East, Madras 600 035, India

First edition 1994

© 1994 Chapman & Hall

Typeset in 10/12 pt Times by Type Study, Scarborough
Printed in Great Britain by Cambridge University Press

ISBN 0 7514 0181 1 ✔

A catalogue record for this book is available from the British Library
Library of Congress Catalog Card Number: 94-70713

∞ Printed on acid-free text paper, manufactured in accordance with ANSI/NISO Z39.48–1992 (Permanence of Paper)

Preface

Wheat provides over 20% of the calories for the world population of 5.3 billion persons. It is widely grown in five of the six continents. It is a highly versatile food product in that it can be stored safely for long periods of time and transported in bulk over long distances. In relative terms, it is reasonably priced; over the past quarter century, the inflation-adjusted price of wheat has been declining. Modern milling and baking technology required for the transformation of wheat grain into consumable baked products is available or accessible in all countries of the world. For these reasons, and because Canada is one of world's leading wheat producing countries, it seemed appropriate to include a major symposium on wheat in the scientific and technical program of the 8th World Congress of Food Science and Technology held in Toronto, Canada during September 29–October 4, 1992.

In selecting the topics for the symposium on wheat, we attempted to cover a full range of subjects including economics and marketing, nutrition, grading, processing, constituent chemistry and functionality, biotechnology, and safety of genetically modified wheat varieties. The major focus was on common hard (bread) wheats; separate papers were devoted to the unique characteristics and technological properties of common soft (biscuit) and durum (pasta) wheats. Each paper was presented by an acknowledged international expert.

This book provides a more permanent record of the papers presented at the symposium. Because of the wide range of disciplines and topics, the chapters vary somewhat in length and in detail, and the personal style of writing of the various authors has been retained as much as possible.

We would like to thank all of the authors for their diligence in preparing and submitting the manuscripts. We would also like to record our appreciation to Heather Delorme for retyping several of the chapters and organizing the materials for publication. We trust that this book will be useful to all who are interested in wheat.

W.B.
V.F.R.

Series foreword

The 8th World Congress of Food Science and Technology, held in Toronto, Canada, in 1991 attracted 1400 delegates representing 76 countries and all five continents. By a special arrangement made by the organisers, many participants from developing countries were able to attend. The congress was therefore a most important international assembly and probably the most representative food science and technology event in that respect ever held. There were over 400 poster presentations in the scientific programme and a high degree of excellence was achieved. As in previous congresses, much of the work reported covered recent research and this will since have been published elsewhere in the scientific literature.

In addition to presentations by individual researchers, a further major part of the scientific programme consisted of invited papers, presented as plenary lectures by some of the leading figures in international food science and technology. They addressed many of the key food issues of the day including advances in food science knowledge and its application in food processing technology. Important aspects of consumer interest and of the environment in terms of a sustainable food industry were also thoroughly covered. The role of food science and technology in helping to bring about progress in the food industries of developing countries was highlighted.

This book is part of a series arising from the congress and including full bibliographical details. The series editors are professor M. A. Tung of the Technical University of Nova Scotia, Halifax, Nova Scotia, Canada; and Dr G. E. Timbers of Agriculture and Agri-Food Canada, Ottawa, Ontario, Canada. The book presents some of the most significant ideas which will carry food science and technology through the nineties and into the new millennium. It is therefore essential reading for anyone interested in the subject, including specialists, students and general readers. IUFoST is extremely grateful to the organisers from the Canadian Institute of Food Science and Technology for putting together a first class scientific programme and we welcome the publication of this book as a permanent record of the keynote papers presented at the congress.

Dr D. E. Hood
(President, International Union of Food Science & Technology)

Contributors

W. BUSHUK Department of Food Science, University of Manitoba, Winnipeg, Manitoba R3T 2N2, Canada.

A. CURIONI Istituto di Biotecnologie Agrarie, Universita di Padova, Via Gradenigo, 6–35100, Padova, Italy.

T. DACHKEVITCH Istituto Sperimentale per la Cerealicoltura, Via Molina, 3–20079, S. Angelo Lodigiano, Italy.

A. DAL BELIN PERUFFO Istituto di Biotecnologie Agrarie, Universita di Padova, Via Gradenigo, 6–35100, Padova, Italy.

B. L. D'APPOLONIA North Dakota State University, Department of Cereal Science and Food Technology, Fargo, North Dakota 58105, USA.

H. FARIDI Nabisco Brands Inc Tech Center, PO Box 1943, East Hanover, New Jersey 07936, USA.

P. FINNEY USDA–ARS, Soft Wheat Quality Lab, Wooster, Ohio 44691, USA.

C. GAINES USDA–ARS, Soft Wheat Quality Lab, Wooster, Ohio 44691, USA.

R. H. KILBORN Grain Research Laboratory, 1404–303 Main Street, Winnipeg, Manitoba R3C 3G8, Canada.

J. E. KRUGER Grain Research Laboratory, 1404–303 Main Street, Winnipeg, Manitoba R3C 3G8, Canada.

A. A. MACDONALD Grain Inspection Division, 900–303 Main Street, Winnipeg, Manitoba R3C 3G8, Canada.

R. R. MATSUO Grain Research Laboratory, 1404–303 Main Street, Winnipeg, Manitoba R3C 3G8, Canada.

F. MEUSER Technische Universität Berlin, Seestraße 11, D-13353 Berlin 65, Germany.

W. R. MORRISON Department of Bioscience & Biotechnology, Food Science Division, University of Strathclyde, 131 Albion Street, Glasgow G1 1SD, Scotland.

B. T. OLESON The Canadian Wheat Board, 423 Main Street, Winnipeg; Manitoba R3C 2P5, Canada.

N. E. POGNA Istituto Sperimentale per la Cerealicoltura, Via Molino, 3–20079, S. Angelo Lodigiano, Italy.

K. R. PRESTON Grain Research Laboratory, 1404–303 Main Street, Winnipeg, Manitoba R3C 3G8, Canada.

G. S. RANHOTRA American Institute of Baking, 1213 Bakers Way, Manhattan, Kansas 66502, USA.

V. F. RASPER Department of Food Science, University of Guelph, Guelph, Ontario N1G 2W1, Canada.

P. RAYAS-DUARTE North Dakota State University, Department of Cereal Science and Food Technology, Fargo, North Dakota 58105, USA.

R. REDAELLI Istituto Sperimentale per la Cerealicoltura, Via Molino, 3–20079, S. Angelo Lodigiano, Italy.

H. J. SCHOCH Buhler Ltd., CH-9240, Uzwil, Switzerland

J. D. SCHOFIELD Department of Food Science and Technology, University of Reading, Whiteknights, PO Box 226, Reading RG6 2AP, UK.

W. SEIBEL Federal Center for Cereal, Potato & Lipid Research, Schuetzenberg 12, D-4930 Detmold, Germany.

K. H. TIPPLES   Grain Research Laboratory, 1404–303 Main Street, Winnipeg, Manitoba R3C 3G8, Canada.

R. TOWNSEND   Pioneer Hi-Bred International Inc., Plant Breeding Division, 7300 NW 62nd Avenue, Johnston, Iowa 50131–0038, USA.

Contents

1 World wheat production, utilization and trade

B. T. OLESON

1.1 Introduction

This chapter outlines the production, utilization and trade background that sets the stage for this book. It begins by setting out a simple framework for the discussion. First, at least from an economic perspective, all things begin with demand. Satisfying the needs of a growing population is the fundamental demand factor in the wheat sector. Second, production systems evolve to satisfy this demand. Third, the potential demand associated with population growth may not be translated into effective demand unless population growth is accompanied by income growth. Finally, the two ecosystems – economics and ecology – must be in harmony.

1.2 Wheat: a brief overview

Wheat is the leading cereal grain produced, consumed and traded in the world today. The cultivation of wheat is thought to have begun several thousand years before the birth of Christ; and bread, leavened and unleavened has been a staple food for humans throughout recorded history. The cultivation of wheat allowed the establishment of permanent settlements, fostering the development of civilization as we know it. Even today, wheat provides more food for people than any other source.

Although it is difficult to determine unequivocally, one United Nations study estimates that 90% of the world's food supplies come from the land. According to this analysis, cereal grains (wheat, corn, rice, barley, sorghum, etc.) provide an incredible 68% of world food supplies. This estimate allows for both direct (bread, pasta, etc.) and indirect (meat, milk, eggs, etc.) consumption of cereals. Other foods such as tubers, fruits and vegetables account for 22%. The remaining 10% of man's food comes from the world's oceans, seas and lakes, in the form of fish and other aquatic life.

Cereal grains are commonly designated as either food (wheat and rice) or feed/coarse grains (corn, barley, sorghum, etc.), based on their primary end-use. Over the past five years (1986 to 1990), world production of all cereal grains has averaged 1.661 billion tonnes. Wheat was the leading grain produced, averaging 533 MMT annually, representing almost one-third of all cereal production. Corn was the second largest grain produced, averaging 451 MMT (or 27%) of world production, followed by rice at 331 MMT (milled basis), accounting for 20%, and barley at 178 MMT (11%). Production of the remaining coarse grains averaged 168 MMT and accounted for only 10% of total world output.

1.3 Wheat classes/grades

The word 'wheat' does not refer to a single species and actually applies to three groups of species which belong to the grass family. Most of the wheat varieties cultivated today are grouped together under the broad category of common or bread wheat, which accounts for approximately 95% of world production. Nearly all of the remaining 5% of cultivated varieties are durum wheats used for such products as pasta and couscous.

Wheat is segregated into various classes according to its agronomic and end-use attributes. These classifications are generally based on quality, color and growth habit. Based on its suitability for baking bread, wheat is normally divided into two quality classes; hard and soft. Hard wheat has a physically hard kernel that yields a flour with high gluten and consequently high protein content and this is suitable for producing a western style loaf of bread and some types of noodles. On the other hand, soft wheat is characterized by a lower protein level and is most suitable for producing cakes and biscuits, which do not require a strong flour. There are also semi-hard wheats having some combination of the above quality character- istics and utilized in unleavened breads such as chapatis as well as Asian steamed bread and certain noodles.

Color (e.g. red or white) refers to the color of the aleurone or outer layer of the wheat kernel. Depending on the end-product and the milling extraction rate, different color wheats may be desirable in different markets. Finally, wheat is also classified according to growth habit, i.e. whether it is a spring or winter wheat.

There can be quite a disparity within the various classes of wheat; so, in order to provide wheat of a consistent quality, it is divided into grades. The grain is graded according to attributes such as test weight, protein content, moisture and foreign material content. The major wheat exporting countries all have distinct grading systems designed to provide wheat that meets a specific standard. Providing wheat of consistent quality from cargo to cargo and from year to year is extremely important to the end-user.

1.4 Wheat production

Wheat has the widest adaptation of all cereal crops and is grown in some 100 countries around the world. It is grown as far north as Finland and as far south as Argentina. The heaviest concentration is in the temperate zone of the northern hemisphere between the 30th and 60th latitudes, which includes the major grain growing areas of North America, Europe, Asia and North Africa. There is also some lesser concentration between the 27th and the 40th latitudes in the south, chiefly Australia, Argentina, Brazil and South Africa.

The growth in world wheat production over the past 30 years is probably one of the most remarkable achievements of the 20th century. Predictions of mass starvation made in the 1960s and 1970s have thus far proved false, with the exception of regions in Africa. Since 1960, world wheat production has been growing faster than population. While world population has almost doubled, from 3 billion in 1960 to 5.3 billion in 1990, wheat production has almost tripled from the 1960 level.

Notwithstanding year-to-year variability, world wheat production has exhibited steady growth from 1960 to 1990, ranging from a low of 225 MMT in 1961 to a record 593 MMT in 1990. On average, wheat production has been growing by over 100 MMT per decade, averaging 273 MMT in the 1960s, 378 MMT in the 1970s, and 494 MMT in the 1980s.

Through the ages production increases have come mainly from increased area; yet from 1960 to 1990, the world wheat harvested area increased only 14% while production increased by 150%. The average harvested area grew by only 4% per decade, from 213 million hectares in the 1960s, to 221 million hectares in the 1970s and 229 million hectares in the 1980s.

Most of the increased wheat production over the past 30 years can be attributed to the success of the 'green revolution.' Since 1960, world wheat yields have more than doubled following the introduction of high-yielding, fertilizer-responsive wheat varieties and better cultural practices. Average wheat yields increased from 1.28 tonnes/hectare in the 1960s to 2.16 tonnes/hectare in the 1980s and a record 2.57 tonnes/hectare in 1990.

Although wheat production is a biological function, it cannot be looked at in isolation from the political and economic factors which play a significant role. Many countries are prepared to go to great lengths to achieve food security. A noteworthy example is Saudi Arabia.

Just ten years ago who would have predicted that the Kingdom of Saudi Arabia, much of which is desert, would become self sufficient in wheat, and much less become an important wheat exporter. In the late 1970s, the Saudi government began looking for ways to invest some of its oil revenues in its developing rural economy. By offering domestic wheat producers a guaranteed price of US$ 1050/tonne, as well as other input subsidies, the Saudi government hoped to increase its food security and reduce its dependence on imported wheat, which amounted to 1.3 MMT in 1970/80.

These incentives proved to be more than enough, and by 1984 the Saudis had achieved self sufficiency, with wheat production of 1.4 MMT. By this time it became apparent that the high price would cause overproduction.

Accordingly, the government dropped the domestic price to US$ 540/tonne, with little effect because returns from wheat were still better than those for alternate crops. By 1990/91, Saudi wheat production had reached 3.6 MMT, and it is still growing. Confronted with a growing surplus, the Saudis began to export wheat in 1986. In 1988/89, Saudi wheat exports were 2 MMT; and last season (1990/91) Saudi Arabia was the sixth largest wheat exporter in the world. Saudi wheat, much of which must be irrigated from non-renewable underground aquifers, has been exported to some 40 countries around the world, at prices a fraction of their domestic price. This includes a recent sale of 21 000 tonnes to New Zealand, which infuriated Australia as well as New Zealand's wheat growers to such an extent that the cargo was eventually diverted on 'high seas' to a less sensitive destination.

1.5   Wheat producers

Looking at the last five years (1986–1990), world wheat production has averaged 533 MMT (including the record 593 MMT produced in 1990 and the drought-reduced 1988 crop of just 501 MMT). Wheat production is highly concentrated. Just three producers, the Soviet Union, China and the EC-12, account for almost half of world production, producing 92 MMT, 90 MMT and 79 MMT, respectively. When the USA (59 MMT), India (48 MMT) and Canada (26 MMT) are added, the top six producers account for about three-quarters of world wheat production. These same six countries held the same aggregate share of world production in the early 1960s; however, in some cases the individual shares have changed quite dramatically.

Although the Soviet Union is still the world's largest wheat producer in relative terms, it has lost considerable ground. From 1960 to 1964, average Soviet wheat production of 65 MMT accounted for 27% of world wheat production. Average (1986–90) production is now 92 MMT, but represents only 17% of world output. Wheat production in the former Soviet Union is highly variable, chiefly due to the vagaries of its climate. Over the past 30 years, Soviet wheat production in the former Soviet Union ranged from a high of 121 MMT in 1978 to only 50 MMT in 1963 and varied by as much as 40 MMT from one year to the next. Moreover, from 1960 the Soviet harvested area for wheat actually declined by over 20%. This can probably be attributed more to political and economic factors, resulting in a declining farm population, than to agricultural factors. In addition, although Soviet wheat yields have improved, they are still only about 75% of world average.

In contrast, wheat output in China has increased dramatically. Chinese wheat production, which averaged 18.3 MMT from 1960 to 1964, increased almost five fold, to an average of 89.6 MMT from 1986 to 1990. At the same time, its share of world wheat production more than doubled, from 8 to

17%. This phenomenal increase is almost exclusively the result of increased yields. Average wheat yield in China over the past 30 years has quadrupled, while the area devoted to wheat has increased by 13% – about world average. In the case of China, increased yields are the result of better varieties and improved cultural practices assisted by the liberalization of grain production and associated economic incentives to produce.

India is one of the leading examples of the 'green revolution.' Since 1960, India's wheat yields have almost tripled due almost exclusively to the introduction of improved varieties and agricultural practices developed as part of the 'green revolution'. At the same time, the number of farmers and the wheat area have almost doubled. India's wheat production of 54 MMT in 1990 was over five times the 1960 level, and its share of world wheat production has more than doubled from 4% in the early 1960s to the current 9%. Consequently, India has been reduced from a major wheat importer to a relatively small importer and occasional exporter. However, in spite of these significant achievements, the challenge for India to feed its growing population remains the single most important issue.

When the European Community was formed in post-war Europe in 1957, one of its primary goals was to achieve self sufficiency in food. Thus, the Common Agricultural Policy (CAP) was formulated, with high support prices designed to stimulate domestic grain production. These high guaranteed prices encouraged intensive high-input farming practices, which substantially increased yields and caused European Community wheat production to more than double, from an average of 37 MMT in the early 1960s, to 79 MMT from 1986 to 1990, and a record 85 MMT in 1990. Over the same period, domestic consumption increased by only 40%. Excess wheat production was exported onto the world market with the use of export subsidies that bridged the gap between the high domestic price and the lower world price. As a result, by 1978/79 the European Community had become a consistent net exporter of wheat and currently (1990/91) holds a 22% share of the world wheat market. European Community overproduction is at the crux of the recent agricultural trade dispute between the European Community and the United States.

1.6 Wheat utilization

Close to 80% of world wheat production is consumed in the country in which it is grown; the balance must be stored or exported. Utilization or disappearance of wheat can be broken down into four categories: food, feed, seed and other (mainly industrial). By definition, over the long term, total world wheat utilization is equal to production. In the short term, utilization exhibits greater year-to-year stability since production changes are smoothed out through the accumulation or release of grain stocks.

Accordingly world wheat utilization has increased from 235 MMT in 1960 to 572 MMT in 1990.

Food is the major use for wheat, accounting on average for two-thirds of total consumption. Over the past ten years, the use of wheat for food has shown quite consistent growth, rising from 298 MMT in 1981/82 to 375 MMT in 1990/91. Wheat is consumed as food in numerous forms, all of which involve some degree of processing. Products such as breakfast cereals generally make use of the whole kernel, but the majority of wheat for food is first milled into flour to be used for bread, noodles, biscuits, cakes, etc. Interestingly enough, in a recent poll conducted on behalf of the Wheat Foods Council in the United States, almost half (49%) of the respondents could not correctly identify white bread as a wheat product.

As a feed grain, wheat is excellent for poultry and is equivalent to corn for many classes of livestock. The use of wheat for feed tends to be highly variable and is dependent on the price relationship between wheat and other feedgrains as well as the quality of the wheat crop in any given year. Feed use of wheat has averaged 104 MMT (or 20%) of total consumption over the past ten years but has varied by as much as 23 MMT from one year to the next, making it the primary cause of variations in world wheat consumption. In 1990/91, due to the unusual price relationship with corn, feed use of wheat soared to a record 121 MMT.

About 7% of wheat utilization is seed. The use of wheat for seed averages about 35 MMT annually and is directly related to the area sown in any given year.

Other, mainly industrial, uses of wheat account for about 6% of disappearance. This includes the processing of wheat into starch and gluten, which have a wide range of food and other uses such as protein enrichment of flour; a binding or strengthening agent in some processed food products; and the production of ethyl alcohol, plastics, varnishes, soaps, rubber, cosmetics, etc.

1.7   Wheat stocks

Despite annual fluctuations, the overall level of world wheat stocks has shown an upward trend from an average of 84 MMT in the 1960s to almost 140 MMT in the 1980s. However, in relative terms (stocks as a percent of utilization), present world stocks are lower than they were in the 1960s. The average stocks-to-use ratio has fallen from 31% in the 1960s to 28% in the 1980s. In spite of the record world wheat crop in 1990/91, carryover stocks of 142 MMT represent only 25% of world use, about a thirteen week supply. However, this is a somewhat misleading notion since wheat harvests are staggered throughout the year, thereby allowing great flexibility for markets to adjust to market shocks both in production and consumption.

As presently compiled, year-end world wheat stocks do not actually represent a particular point in time, but rather are an aggregate of the carryover stocks of the various importing and exporting countries often determined at different times. As such, world stock levels represent more of a trend than an actual figure. Changes in world wheat carryover stocks have traditionally been used as a barometer of potential market volatility. When world wheat stocks are low relative to utilization, a production shortfall or unusually high demand anywhere in the world could potentially have a major impact on world supplies and prices. Naturally this is an over simplification. It is important to know where stocks are being held, what the stock level of the various classes of wheat is, whether the stocks are readily accessible to the market, etc.

A recent International Wheat Council (IWC) report suggests that, although stocks (particularly those of the major exporters) are still a key determinant of price, the relationship between stocks and prices has been weakened. Grain exporting countries are less likely to be surprised by unforeseen demand because, in many countries, credit is a limiting factor. At the same time, the market may be less susceptible to successive crop failures in view of acreage set aside in the United States and the improved position of exporting countries to respond to changes in demand due to their improved grain handling and transportation facilities.

1.8 World wheat trade

Wheat is the most important cereal grain traded in the world today, both in terms of absolute volume (97 MMT [1986–90 average]) and share of world trade in all cereal grains (49%). Coarse grain (mainly corn) exports of 89 MMT account for 45% of world trade, while only 13 MMT of rice are traded annually (representing 6% of world trade).

Wheat is also the cereal grain most dependent on the export market, with almost one-fifth of wheat production exported. This compares with 11% of coarse grain production and only 4% of rice entering export channels.

Over the past 30 years, world wheat trade has ranged from 42 MMT in 1960/61 to a record 107 MMT in 1984/85. World wheat trade tends to be quite volatile, and this volatility is increasing. In the decade of the 1960s, wheat trade averaged 50 MMT, while year-to-year imports varied by as much as 12 MMT. By the 1970s, average world wheat trade had grown to 66 MMT, with imports varying by as much as 15 MMT from one year to the next. In the 1980s, trade increased to 98 MMT, while varying by as much as 22 MMT from one year to the next.

World wheat exports are highly concentrated, with over 90% of wheat trade conducted by only four countries and the EC-12. From 1986 to 1990, the United States was the leading exporter, with 35% of world trade.

Canada was second, with 20%, followed closely by the EC-12, with 19%. Australia and Argentina were in fourth and fifth place, with 12% and 5%, respectively.

Over the past 30 years, there has been a major shift in world wheat imports from the developed to the developing countries. In 1960/61, 43% of wheat exports went to Europe, and only 14% was destined for Africa and the Middle East. The former Soviet Union was a net exporter of 5 MMT of wheat. By 1990/91, Europe's share of world wheat imports had fallen to only 5%, Africa and the Middle East's share had more than doubled to 33%, and Soviet imports represented 14% of world trade.

Wheat imports are also concentrated. The top five wheat importing countries account for almost half (47%) of world trade. At the same time, almost every country in the world imports some quantity of wheat or flour.

Two countries, the Soviet Union and China, together account for 30% of world wheat imports. Over the past five years, Soviet wheat imports averaged 16.4 MMT, or 17% of world trade, while Chinese imports were 12.3 MMT, accounting for 13%. Because they are also the world's top two wheat producers, imports by these countries can vary considerably from year to year, having a major impact on world trade and prices. In the past ten years, Soviet wheat imports have varied by as much as 12 MMT from one year to the next, while Chinese imports have varied by 6 MMT. Egypt is the third largest importer, with 6.6 MMT (7%), followed by Japan, with 5.6 MMT (6%) and Algeria, with 4.1 MMT (4%). Other significant importers include Iran (3.8 MMT), South Korea (3.5 MMT) and Iraq (2.5 MMT).

The activities of the former USSR, the world's largest wheat importer, have had a major impact on the world wheat market. It is interesting to note that at the beginning of each decade the grain industry has been shocked by wheat trade developments associated with the former Soviet Union. In the 1960s, the USSR surprised the world with colossal wheat imports; the 1970s brought us the 'Great Grain Robbery'; in the 1980s we were jolted by the United States-led grain embargo; and now, in the 1990s, we are trying to understand the impact of the new Russian Revolution on the grain trade.

1.9 Wheat prices

Over the past century, the inflation-adjusted price of wheat has been trending downward. In the past few years, the price war and the escalating use of export subsidies has led to a virtual collapse of world wheat prices. There have actually been two phases in the current price war. The first was the period prior to the summer of 1988, at which time severe drought interrupted hostilities, and the second was the period 1989 onward, during which time hostilities resumed.

A brief reference to market shares of the United States and the European Community summarizes key points regarding the current trade war. Special factors led to the United States capturing almost half of world wheat trade in 1973/74. However, their market share fell by the early to mid-1980s to about one-third. The Americans attributed this lower market share to the subsidized competition from the European Community. In 1985, the United States introduced BICEP, now the Export Enhancement Program (EEP), to subsidize United States exporters to compete against the European Community in certain 'targeted' markets by offering export bonuses or subsidies. Since the introduction of EEP, the world wheat market has effectively been operating on two levels: a 'commercial,' or unsubsidized level, and a 'subsidized' level.

With the exception of the drought years of 1988 and 1989, commercial wheat prices as represented by the price of US HWORD FOB Gulf have been declining over the past decade. In 1980/81, the average price for US HWORD out of the Gulf was US$ 182/tonne. In 1990/91, average commercial wheat prices fell to only US$ 118/tonne. In addition, a substantial portion of wheat was traded at subsidized prices (determined by subtracting the EEP bonus from the HWORD price), which fell from about US$ 116/tonne in July 1990 to only US$ 61/tonne by August 1991. Not only have EEP-subsidized wheat prices fallen, the share of United States exports that are subsidized under EEP has increased from 12% in 1985/86 to more than half of all exports in 1990/91. The European Community has kept pace with these United States initiatives by increasing its export subsidies accordingly. In February 1991, the European Community export subsidy was US$ 192/tonne while the export price was only US$ 74/tonne.

Fundamental factors also played a role in the fall of world wheat prices. Back-to-back record wheat crops of 538 MMT in 1989 and 593 MMT in 1990 created ample supplies of wheat. Wheat production increased in all five major exporting countries as well as the two leading importing countries. In the battle to maintain market share in a shrinking market, export subsidies have been raised to record levels while world wheat prices (inflation adjusted) have fallen to historically low levels.

While most wheat traded is destined for human consumption, the fact that wheat is an excellent animal feed means that wheat prices on the down side cannot be viewed in isolation from other grain prices, particularly corn. It is astounding to note that in September 1991, subsidized wheat was trading at almost half the price of corn. In fact, at a recent seminar sponsored by *Milling & Baking News*, United States bakers were told that 'the single most important influence on flour costs in 1991/92 will be the per-acre yield of this year's corn crop'.

1.10   Long term factors

In the longer term, the factors which could influence world wheat production and trade are endless. There is really no way to predict or fully control the ultimate outcome of the interplay of the various climatic, political and economic factors which will determine future production and trade patterns.

Even the outcome of a relatively stable variable, like population growth, is not certain. World population is currently growing at an annual rate of 1.8%, which translates into an extra 94 million people per year. The United Nations offers a range of projections as to when the world's population will reach replacement fertility and thus level off. Based on the median assumption, fertility rates would level off in the year 2035, and population would stabilize at about 10 billion people by the end of the 21st century. This is still subject to a wide variability. The difference between the high and low extremes is 5 billion people. In addition, population growth is not evenly distributed. While population in developed countries has stabilized, virtually all future growth will take place in developing countries. The growth in potential demand will be great, but will this potential be translated into effective demand which is so dependent on ability to pay?

Environmental concerns are becoming a real issue. There are already indications that the high-input agriculture practiced in the European Community, with extremely high application of chemicals such as nitrogen, is having a major adverse effect on the environment. At the extreme, the application of nitrogen fertilizer in the Netherlands is 20 to 30 times the average on the Canadian Prairies. The government there is working to reduce excess nitrogen accumulation to near zero by the year 2000, and fines have already been legislated for violators. It is likely that the agricultural scientists will place environmental concerns as a high priority in their research efforts in the 1990s.

1.11   Concluding comment

> The nation with too much bread has many problems,
> the nation with too little bread has only one problem.
> *Fifth Century Byzantine Proverb*

The above factors obviously pose a major challenge to the scientific community. How best do we balance the needs of a growing world population against the limits to growth that may be imposed by the environmental considerations of spaceship earth? Whatever the pressures, whatever the solutions, one thing is certain: wheat will continue to be center stage in world efforts to feed a growing population. With trade, we can escape the confines of the nation in the Byzantine proverb. As a globe, we cannot.

References

Avery, D. T. *Global Food Progress 1991*, Hudson Institute, Indianapolis, 1991.
Canada Grains Council. *Wheats of the World*, Canada Grains Council, Winnipeg.
Dommen, A. J. (1989) *World Agriculture Situation and Outlook*, Agriculture and Trade Analysis Division, Economic Research Service, US Department of Agriculture, March.
Heyne, E. G. (ed.) (1987) *Wheat and Wheat Improvement*, 2nd edn, American Society of Agronomy.

2 Wheat: contribution to world food supply and human nutrition

G. S. RANHOTRA

2.1 Introduction

Wheat is the leading cereal grain produced in the world, followed closely by rice and corn (maize). The other major cereals are sorghum, barley, oats, rye and millets. Wheat is grown to some extent on every continent except Antarctica.

It is often difficult to obtain accurate information on the quantity of wheat grown worldwide because of the continually changing conditions and the lack of reliability of the statistics reported by many governments. Nevertheless, estimates are routinely made. The International Wheat Council estimated wheat production in 1991–92 to be 565 million metric tons. This production was second only to the 1990–91 record of 597 million tons. The four largest producers were China, the former USSR, the 12-nation European Community and the United States. The largest exporters in 1988–89 were (in order of decreasing amounts shipped) the United States, the European Community, Canada and Australia. (The International Wheat Council estimates wheat production in 1993–94 to be 547 million metric tons, or about 92% of the 1990–91 record.)

2.2 Utilization of wheat

Wheat is closely associated with human food uses. It is estimated that nearly two-thirds of the wheat produced in the world is used for food; the remaining one-third is used for feed, seed and non-food applications.

In many countries, wheat is the major component of the diet. it is non-perishable, easy to store and transport, has a good nutritional profile and allows the manufacture of a wide variety of interesting, enjoyable and satisfying products. These products require flour of select characteristics which are achieved through a proper balance of grain hardness and protein content (Figures 2.1 and 2.2).[1,2]

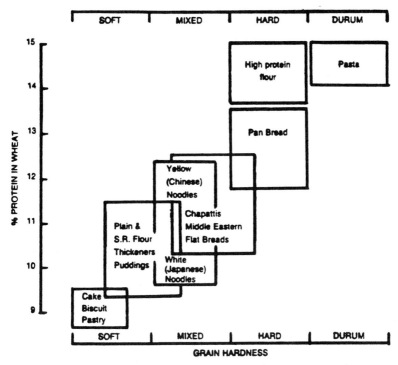

Figure 2.1 Wheat types for various end products (S.R. = self raising).

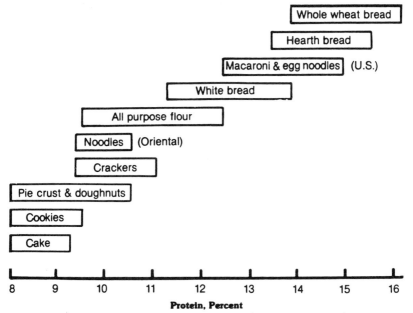

Figure 2.2 Percentages of protein in wheat used in different end-products.

Table 2.1 Total daily calories contributed by wheat foods (United States)

Food	Men (%)	Women (%)
Bread and related products	9.1	8.3
Breakfast cereals	0.9	1.1
Crackers	0.6	0.8
Pastas	2.4	2.2
Sweet goods	5.3	4.8
Total	18.3	17.2

2.3   Nutritional profile

For many, wheat-based foods are the major source of energy, protein and various vitamins and minerals. In some population groups, wheat-based foods provide two-thirds or more of the daily caloric intake.

2.3.1   Caloric contribution

In the North American diet, wheat-based foods provide nearly one-fifth of the total calories (Table 2.1).[3] At the turn of the century, wheat foods provided substantially more calories when flour consumption per capita per year averaged well over 200 pounds (lb). A steady decline in flour consumption has occurred since that time (in the 1980s, flour consumption averaged between 110–120 lb per capita per year). A reversal of this trend has been noticed in recent years. Flour consumption in the United States today stands close to 130 lb per capita per year. This may be a consequence, in part, of the various dietary recommendations that have been made in recent years to include more complex carbohydrates in the diet. The 1977 Dietary Goals Report[4] recommended that Americans double their calories from complex carbohydrates (Table 2.2)

Table 2.2 Dietary goals for the United States

Calories from	Current diet (%)	Dietary goals (%)
Fat	42	30
saturated	16	10
unsaturated	26	20
Protein	12	12
Carbohydrates	46	58
complex	22	40–45
simple	24	15

Table 2.3 Weight loss on bread-based diets

Bread type	Body weight (kg)		Blood cholesterol (mg dl^{-1})	
	Initial	Loss	Initial	Loss
Regular	90.8±4.5	6.3±4.4	231±11	155±12
High fiber	94.0±5.8	8.8±5.2	224±21	172±12

Although caloric inadequacy is a major health concern in many developing countries, overconsumption of calories, and the consequent overweight in the individual, is more of a problem in the Western world. Wheat-based products are usually the first casualty when embarking on a weight reducing diet. It does not have to be this way. Not only are wheat-based products such as bread and pasta not high in calories, but they can be of assistance when trying to lose weight. The feeling of fullness these products create can help a dieter control food intake. Restricting caloric intake at the same time can also cause a substantial weight loss. Studies conducted in the United States,[5] Germany[6] and Australia[7] have all shown such to be the case when bread was used as the main staple. In the United States' study, overweight college-age men who consumed 12 slices either of high-fiber (reduced calorie) or of regular white bread registered a significant weight loss in eight weeks (Table 2.3). They also showed an appreciable drop in blood cholesterol levels.

2.3.2 Protein contribution

A diet adequate in calories often ensures protein adequacy. However, even when protein adequacy is not a concern, wheat alone will not provide all the essential amino acids in the amounts needed for proper growth and maintenance of good health. As in all other grains, lysine is the most deficient amino acid in wheat. The amino acid profile of wheat improves dramatically, however, when only a small quantity of leguminous or animal protein is simultaneously included in the diet, or a product is made with flour blended with non-wheat protein sources. In a recent study,[8] Balady bread made with wheat flour (extraction 87%) blended with rice flour and corn flour and containing either non-fat dry milk or full-fat soy flour showed an impressive improvement in lysine content and protein quality (Table 2.4).

In Western countries, protein consumption is often more than adequate. In the United States, protein intake averages over 100 g per day. This roughly represents two to three times the amount needed. About two-thirds of this protein originates from animal sources. These patterns of protein intake obviate the need to fortify foods with protein sources. Where still used, protein and amino acids are added strictly for functional reasons,

Table 2.4 Protein quality of Balady bread made with various flour blends

	Blend/Bread					
	A	B	C	D	E	F
Blend composition (%)						
Wheat flour	100	70	70	40	40	40
Rice flour	–	30	–	30	25	25
Corn flour	–	–	30	30	25	25
Dry milk	–	–	–	–	10	–
Soy flour	–	–	–	–	–	10
Bread composition						
Protein (%)	10.3	9.4	9.4	8.4	10.0	10.2
Calories (per 100 g bread)	269	276	274	284	279	279
Lysine (g/100 g protein)	2.1	2.0	2.0	2.4	3.2	3.1
Chemical score[a]	32.8	30.5	31.0	39.3	51.3	48.8
Protein efficiency ratio[b]	1.0	1.0	1.0	1.1	1.4	1.6

[a] Based on FAO/WHO lysine value of 5.5 g/100 g protein. [b] Based on casein value of 2.5

e.g. gluten added to variety breads to maintain quality and cysteine added to the dough to reduce mixing time.

2.3.3 Fat in wheat-based foods

Wheat itself is low in fat, and the fat that is present is high in unsaturated fatty acids which lower elevated blood cholesterol levels, a risk factor in heart disease. Bread, pasta products and many wheat-based breakfast cereals are low in fat. In contrast, sweet goods are usually high in fat, but animal fats are now rarely used in these products produced in North

Table 2.5 Low-fat yellow cake

Ingredient	True percentage
Cake flour	21.61
Sugar	18.29
Fat substitutes	
Litesse™	7.00
N-Flate™	4.09
Xanthan Gum	0.10
Gluco delta lactone	0.20
Liquid whole eggs	17.77
Baking soda	0.16
Baking powder	1.83
Flavors (butter, vanilla)	0.16
Salt	0.35
Water (for dry eggs)	28.44

Fat content: 1.3% (4.8% by acid hydrolysis)

Table 2.6 Total and soluble fiber in wheat and its fractions

	Total dietary fiber (%)	Soluble fiber (%)
Straight flour	2.5	1.1
Patent flour	2.5	1.3
Wholewheat flour	10.2	1.3
Germ	9.3	1.1
Bran	44.0	2.1

America; most products contain oil or shortenings made from vegetable oils which retain some degree of polyunsaturation and, of course, contain no cholesterol.

In North America, sweet goods are now also being modified to be low or free of fat and contain little or no cholesterol. Several fat substitutes used in this reformulation are based on starches obtained from wheat and other sources. Table 2.5 gives an example of low-fat yellow cake which used three fat substitutes.

2.3.4 Dietary fiber

Wholewheat flour and its bran fraction are a good source of fiber, particularly water insoluble fiber (Table 2.6).[9] In contrast, white flour, although not high in total fiber, is relatively high in soluble fiber. While soluble fiber may help normalize elevated blood cholesterol and sugar levels, the insoluble fraction may be helpful in the prevention and management of several disorders of the intestinal tract.

Wheat bran promotes fecal bulk and provides regularity (coarse bran is more effective than ground bran). This bulking effect may also reduce the risk of colorectal cancer, a major health problem in North America. Table 2.7 outlines the probable mechanisms through which wheat bran and other

Table 2.7 Carcinogenesis hypothesis

Fiber may favorably alter colonic microflora such that carcinogenic compounds are not produced.

Fiber reduces fecal transit time. This reduces the contact between carcinogens and mucosal cells.

Fiber increases fecal bulk. This would dilute the concentration of carcinogens.

Fiber stimulates bacterial proliferation in the colon. This uses up ammonia which is reported to promote carcinogenesis.

Fiber produces butyric acid. This acid is reported to reduce the liability of cells to malignant change.

Table 2.8 Composition of diets

	Control diet	Bakery diet
Percent calories from:		
Fat	30	30
Protein	15	15
Carbohydrates	55	55
Cholesterol intake (mg/day)	450	450
Fiber intake (g/day)		
Total fiber	25	25
Soluble fiber	3	9

fecal bulking agents may reduce the incidence of colon cancer. A high fiber diet may also help prevent breast cancer, probably through effective changes in the hormone system.

Soluble fiber in oats, beans, pectins, etc. has been shown to lower serum cholesterol in human subjects. In animals, soluble fiber in bread products has also been shown[10] to have a similar effect. To confirm this in human subjects, a clinical study was recently undertaken.[11] Thirteen male subjects with primary hypercholesterolemia (serum cholesterol, 200–320 mg dl^{-1}) were recruited for the 4-week study. Ten completed the study. No subjects were receiving cholesterol-lowering medications. These subjects were admitted to a metabolic ward for a 7-day baseline period during which subjects adjusted to a diet high in fiber but low in soluble fiber. Subjects then received a high soluble fiber diet emphasizing refined, wheat-based products for 21 days.

Both the control and bakery diets provided near identical levels of carbohydrates, fat, protein, cholesterol and total fiber (Table 2.8). The control and bakery diets, however, differed in the amount of soluble fiber. Bakery products in the control diet were selected to provide primarily insoluble fiber (control diet contained only 3 g soluble fiber); the bakery diet contained 6 g additional soluble fiber which originated primarily from breads, buns, waffles, crackers and biscuits. Neither diet contained oats or legumes (some soluble fiber originated from fruits and vegetables). Results obtained are summarized in Table 2.9.

Table 2.9 Serum lipid responses in men

Lipid	Control diet	Bakery diet	Percentage change
Total cholesterol (mg dl^{-1})	223±9	208±8	−6.4
Low-density-lipoprotein cholesterol (mg dl^{-1})	149±7	135±5	−8.5[a]
High-density-lipoprotein cholesterol (mg dl^{-1})	30±1	29±1	−3.0
Triglycerides (mmol l^{-1})	2.46±0.24	2.46±0.27	+0.6

[a] $P < 0.05$ versus control

Table 2.10 Reduced calorie[a] white bread

Composition	White bread	High fiber bread[b]
Moisture (%)	38.0	46.0
Protein (%)	7.5	9.5
Fat (%)	2.9	3.0
Ash (%)	2.0	2.5
Total fiber (%)	3.0	15.0
Carbohydrates (%)	46.6	24.0
Calories (per 100 g bread)	243	161

[a] Defined as an at least 25% reduction; 'light in calories' is an at least one-third reduction.
[b] Contained added vital wheat gluten.

A wide selection of processed fiber sources is now available. Their use can not only increase the fiber content of a food product, but it can also lower its caloric content. A high fiber source (fiber, over 70%) replacing flour at a 30% level can effectively reduce the caloric content of bread by one-third (Table 2.10).

2.3.5 Vitamins and minerals

Wholewheat flour contains more micronutrients than white flour; the magnitude of this difference widens as the extraction rate of the flour decreases (Figure 2.3).

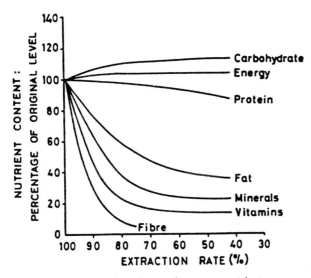

Figure 2.3 Effect of extraction rate on nutrients.

Table 2.11 Hydrolysis of phytate in soy-fortified wheat bread

	Compressed yeast (g lb^{-1} loaf)			
	0	3	9	15
Phytic acid phosphorus				
In bread (mg/loaf)	167	78	37	42
Hydrolyzed (%)	46	75	88	86
Inorganic phosphorus				
In bread (mg/loaf)	185	226	215	231
Increase (%)[a]	440	539	513	551

[a] Over values in bread ingredients.

Wholewheat flour, however, is also high in certain inhibitors of nutrient absorption such as fiber and phytates. Although the level of fiber changes little during food processing, the phytate level can decrease substantially, thus negating the adverse effect of phytates. Phytases in wheat and yeast can bring about a sizable hydrolysis of phytate during breadmaking (Table 2.11).[12] Even when processing may not diminish the phytate level, the higher levels of minerals and vitamins in wholewheat flour (as compared to white flour) may still mean a greater net absorption of these nutrients.

Minimally processed wheat foods have gained popularity in several countries. However, many continue to prefer products made with white

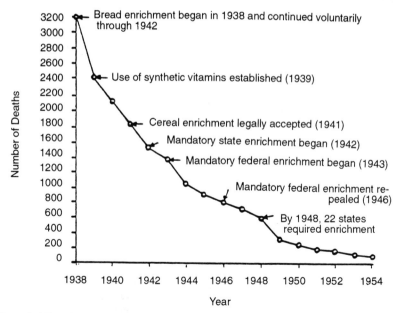

Figure 2.4 Death from pellagra in the United States (1938–1954). Source: National Center for Health Statistics.

Table 2.12 Contribution of cereal fortification towards nutrient intakes

Nutrient	Contribution (%)	
	Through natural occurrence	Through fortification
Thiamin	12	30
Riboflavin	10	18
Niacin	11	19
Iron	15	18

flour such as white bread, pasta and tortillas. Accordingly, white flour has preferably been chosen to enrich (fortify, nutrify) foods to correct mineral and vitamin deficiencies in many societies, including the United States.

Enrichment of flour initiated in the United States in the mid 1940s has been quite effective in virtually eliminating deficiency diseases such as beriberi, ariboflavinosis and pellagra. Figure 2.4 illustrates this quite convincingly for the disease condition pellagra. This effectiveness was the critical consideration in the 1990 Nutrition Labeling and Education Act (effective May 8, 1994) to eliminate, except under certain conditions, mandatory listing of B vitamins on nutrition label now called 'Nutrition Facts'. Deficiency of these vitamins is still a major health concern in many countries, however.

The enrichment program, carried out in the United States since the 1940s, contributes more of the select nutrients towards our need than the amounts contributed through natural occurrence (Table 2.12).[13] Altogether, this contribution is quite substantial. For example, a serving (5.8 ounces) of enriched pasta (cooked) provides one-fourth of the daily allowance (expressed as United States Recommended Daily Allowance) of thiamin, one-fifth of niacin, one-seventh of iron and one-tenth of riboflavin; it also makes a significant contribution to various other nutrients (Figure 2.5).[14] When salt is not used during the cooking of pasta, cooked pasta is virtually free of sodium.

Sodium restriction is now widely advocated for individuals who are hypertensive and sensitive to sodium. Over 60 million Americans show hypertension (elevated blood pressure). Several million of these are sensitive to sodium. According to the 1989 edition of the Recommended Dietary Allowance, our daily need for sodium rarely exceeds 0.5 g per day.[15] On the average, Americans consume 5–7 g of sodium per day.

Wheat is quite low in sodium. However, many wheat foods are high in sodium which is invariably added during food processing such as bread-making and pizzamaking (Table 2.13).[16] Various wheat-based products are now being formulated to be low in sodium, and they are finding acceptance in the market place.

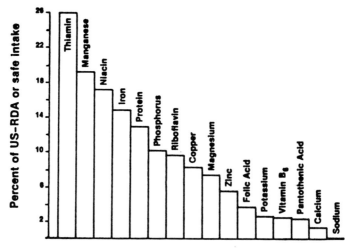

Figure 2.5 Contribution of one serving (165 g) of cooked pasta (spaghetti, macaroni and egg noodles) as percentage of US-RDA or minimum safe adequate intakes for various nutrients.

Table 2.13 Sodium content of pizza

Pizza type	Total sodium (%)		Total (mg/100 g)
	In crust	In non-crust portion	
Cheese	58±5	42±5	524± 72
Hamburger	49±8	51±4	611± 70
Bacon	46±4	54±3	679±104
Pepperoni	40±2	52±4	742± 52
Combination	48±8	52±8	609± 47

While the elimination of B vitamin deficiencies attest to the effectiveness of enrichment programs in the United States, deficiency of iron and calcium continues to be a public health concern. This is, in part, because iron sources added to flour are less well absorbed than iron in meat products (heme iron) and that calcium, although an optional enrichment nutrient in the United States, was rarely added to flour. This situation is changing. The bioavailability of iron added to flour has now been greatly improved (Table 2.14)[17] and enrichment of flour with calcium has also begun.

Calcium not only may help delay an early onset of the osteoporotic disease process, it may also protect against colon cancer and hypertension. Wheat flour normally contains about 200 parts per million (ppm) calcium. Fortification in the United States allows the addition of 2115 ppm, thus yielding a total calcium level in the flour of about 2315 ppm. At the current flour consumption level (150 g per day), calcium enriched flour would provide over 300 mg calcium.

Table 2.14 Iron sources used in cereal fortification

Source	Iron content (%)	Cost ($) per[a] (kg)	Cost ($) per[a] (kg iron)	Bioavailability[b]	Color	Stability
Reduced iron (reduced by hydrogen)	98	1.94	2.00	Fair	Black	Good
Electrolytic iron (reduced iron)	98	4.71	4.80	Fair	Black	Good
Ferrous sulfate[c]	32	2.35	7.30	Good	Tan	Fair
Ferric orthophosphate	28	2.73	9.80	Poor	White	Excellent
Ferrous fumerate	33	2.94	8.90	Good	Red	Fair
Iron EDTA[d]	13	6.60	50.90	Excellent	White	Good

[a] 1990 cost. [b] How well it is absorbed by humans. [c] Now used in pasta products (in place of ferric orthophosphate). [d] Cost of agriculture grade source may be lower.

2.4 Conclusions

Societies consuming diets high in grain-based foods show a lower incidence of chronic degenerative diseases such as heart disease, colon cancer and diabetes.

Even if this consideration is disregarded, it is well to remember that grain-based foods are economical foods as compared to most other foods.

Grain-based foods also allow conserving energy since feeding grains to animals to produce meat is a highly inefficient process.

To align dietary recommendations with food groups, nutritionists in the United States have developed a pyramid concept (grain-based foods occupy the base of the pyramid) to replace the Basic Four Food Wheel concept that gave all four food groups equal billing.

In societies where nutrient deficiencies are still a major health problem, grain-based foods provide an effective vehicle to fortify foods.

References

1. Moss, H. J. (1973) Quality standards for wheat varieties. *J. Aust. Inst. Agri. Sci.*, **39**, 109–115.
2. Eustache, D. (1988) Wheat flour milling. *Technical Bulletin, American Institute of Baking*, **10(11)**, 1–4.
3. Gustafson, N. J. (1983) Consumption and nutritional contribution of wheat foods: Perception and Reality. *Cereal Foods World*, **28**, 631–633.
4. United States Senate (Select Committee on Nutrition and Human Needs). (1977) *Dietary Goals for the United States*, US Government Printing Office, Washington, DC, pp. 1–79.
5. Mickelson, O., Makdani, D. D., Cotton, R. H., Titcomb, S. T., Colmey, J. C. and Gatty, R. (1979) Effects of a high fiber bread diet on weight loss in college-age males. *Am. J. Clin. Nutrition*, **32**, 1703–1708.
6. Stellar, W. (1979) A balance reducing diet based on bread. *Bakers Digest*, **10**, 20–23.

7. Simons, L. A., Stone, P. L. and Simons, J. (1986) The use of bread in diets for weight reduction. *Food Tech. Australia*, **38(2)**, 68–70.

8. Domah, M. B., Ranhotra, G. S., Gelroth, J. A., Glaser, B. K. and Eisenbraun, G. J. (1994) Quality and nutritional characteristics of Balady bread made with various flour blends. *Cereal Chem.*, unpublished data.

9. Ranhotra, G. S., Gelroth, J. A. and Astroth, K. (1990) Total and soluble fiber in selected bakery and other cereal products. *Cereal Chem.*, **67**, 499–501.

10. Ranhotra, G., Gelroth, J. and Bright, P. (1987) Effect of the source of fiber in bread-based diets on blood and liver lipids in rats. *J. Food Sci.*, **52**, 1420–1422.

11. Anderson, J. W., Lawrence-Riddell, S., Floore, T. L., Dillon, D. W. and Oeltgen, P. R. (1991) Bakery products lower serum cholesterol concentrations in hypercholesterolemic men. *Am. J. Clin. Nutr.*, **54**, 836–840.

12. Ranhotra, G. S., Loewe, R. J. and Puyat, L. V. (1974) Phytic acid in soy and its hydrolysis during breadmaking. *J. Food. Sci.*, **39**, 1023–1025.

13. Cook, D. A. and Welsh, S. O. (1987) The effect of enriched and fortified grain products on nutrient intake. *Cereal Foods World*, **32**, 191–196.

14. Ranhotra, G. S., Gelroth, J. A., and Novak, F. A. (1986) Nutrient profile of pasta products commerically produced in the United States. *Nutr. Rep. Internat.*, **33**, 761–768.

15. National Research Council. (1989) *Recommended Dietary Allowances*, 10th edn., National Academy of Sciences, Washington, DC, pp. 247–255.

16. Ranhotra, G. S., Vetter, J. L., Gelroth, J. A., and Novak, F. A. (1983) Sodium in commercially produced frozen pizzas. *Cereal Chem.*, **60**, 325–326.

17. Ranhotra, G. Adding Iron to Food. (1991) *Technical Bulletin, American Institute of Baking*, **13(2)**, 1–6.

3 Bread-wheat quality defined

K. H. TIPPLES, R. H. KILBORN and
K. R. PRESTON

3.1 Introduction

The title of this chapter implies, perhaps, that the reader will receive a clear-cut definition of wheat that is eminently suitable for making bread. However, the subject does not lend itself to simple black and white statements. It depends on what we mean by 'bread'; it depends on what we mean by 'quality'; and it depends on who is doing the defining.

This chapter is divided into several sections as follows:

1. A brief discussion of 'bread' from the point of view of type of product and the various processes used by bakers to produce these products.
2. A discussion of 'wheat quality,' in general terms, to indicate what factors characterize wheat suitable for breadmaking as opposed to wheat that is unsuitable for making bread.
3. A description of some of the methodology and approaches used by our laboratory to evaluate bread quality and bread-wheat quality.
4. A discussion of 'bread-wheat quality' from several different points of view including those of the wheat exporter, the flour miller, the baker and the consumer.

3.2 Bread

To say that breadmaking technology is a complicated subject is an understatement. Not only are there many different shapes and sizes of loaf and variations in crust and crumb characteristics, but there is an infinite number of combinations of formula and processing conditions to be considered. Even within any one country 'bread' may cover a wide range of products. Not only may the type of flour used be quite different, ranging from white flour, free of germ and bran, to stoneground wholewheat, but, on a global basis, we can see flat breads, hearth or free standing loaves, bread baked in open or closed tins, loaves with light or dark crusts and with close or very open crumb structure, loaves of high specific volume or dense loaves of low specific volume. The variations are endless.

If we next consider the many ways bread is made, we must consider questions of formula and processing. The minimum basic formula for leavened bread includes flour, water, salt and yeast or some other source of leavening. In many countries, and especially in North America and Japan, consumers have a taste for sweet, short-eating products containing sugar, fat, milk powder and other such 'rich formula' ingredients. Bakers are also in the habit of using a wide range of additives or improvers to increase dough processing quality, loaf volume, loaf appearance, crumb softness and keeping quality. Such ingredients include various enzymes, emulsifiers, oxidants, etc.

When it comes to processing variations used by bakers, there are, again, many. The basic method may vary from so-called 'straight dough methods' with varying periods of fermentation time, to sponge and dough, liquid ferment and short systems involving high-speed mixing, such as the Chorleywood bread process. The different steps such as mixing, fermentation, sheeting and moulding, proofing and baking are themselves variable in terms of specific conditions such as time and temperature. Furthermore, they may be carried out by hand or by machine and on a small or large scale.

Dough mixing, for example, may be carried out by hand in the most primitive instances, or in dough mixers that vary widely in capacity, mixing action and mixing intensity.

3.3 Wheat quality

Cereal scientists and technologists have been trying for many years to define, understand and measure wheat quality. Wheat is unique among cereal grains in that flour milled from it is capable of making a light, palatable, well-risen loaf of bread when processed into fermented dough. It is the insoluble storage proteins of wheat, collectively called 'gluten', that give wheat its functional properties. Gluten has been described as the most complicated rheological material known to man. (Rheology is the study of viscoelastic materials: gluten has some of the properties of viscous materials, notably flow, and some of the properties of elastic materials which stretch under stress and spring back when tension is relaxed.) It is chemically complex because it is made up of hundreds of different protein subunits which aggregate to form extremely large molecules when made into dough. The presence or absence of specific proteins may well determine whether a variety has 'good' or 'poor' baking quality or whether it is 'weak' or 'strong.' It is gluten that gives dough its ability to form thin sheets that will stretch and hold gas. Even relatively poor quality wheat can produce bread that is significantly more palatable than that made from flour from other cereal grains.

In order to become more discerning and match suitability for bread-making with wheat quality, we should first consider the three basic quality factors that determine wheat type. These are hardness, gluten strength and protein content. In general, soft wheats with weak gluten and low protein content are not suitable for breadmaking; hard wheats with strong gluten and high-protein content are preferred.

3.3.1 Hardness

Because hard wheats require more grinding energy to reduce endosperm chunks into flour-sized particles, a considerable number of starch granules become physically damaged during the milling process. Soft wheats, by contrast, produce flours with low levels of damaged starch. Since damaged starch granules absorb more water than intact, undamaged granules, water absorption is normally significantly higher for hard wheat flours than for soft wheat flours at an equivalent level of protein.

Water absorption is an important quality factor to the baker because it is related directly to the amount of bread he can produce from a given weight of flour. It also has a profound influence on crumb softness and bread-keeping characteristics.

3.3.2 Gluten strength

Doughs made from 'strong' flours tend to be less viscous and more elastic than doughs from weak flours, but generally require more mixing time and energy in order to reach an optimum rheological state. Properly developed doughs have the ability to stretch and form thin sheets that will hold gas and promote production of light, well-risen loaves.

Dough-handling properties during processing may range from soft and sticky for doughs from weak flours to extremely inelastic and tough for doughs produced from flours that are overly strong. Well-balanced dough properties that fall between these two extremes are required for baking. Mixing requirements must also not be too short or too long.

3.3.3 Protein content

Protein content is important in breadmaking because, other things being equal, high-protein flours have a high loaf volume potential, high water absorption and produce loaves with good keeping quality. A fairly wide range of protein content may be incorporated in bread flour grists, but wheats of less than 11% protein are usually unsuitable for breadmaking if used alone. These type-determining factors are generally controlled genetically by variety although protein content may vary widely for a given variety according to location, soil fertility, rainfall, etc. Due largely to genetic selection, wheats of high protein content tend to be hard, to have strong gluten and to produce good bread, whereas wheats of low-protein content tend to be soft, to have weak gluten and to produce small loaves of inferior crumb structure.

3.3.4 Soundness

Apart from the three major type-determining factors, one important wheat quality factor influencing baking quality is soundness. Sound wheat contains very low levels of an enzyme (α-amylase) that attacks and liquifies starch. Germination is associated with a rapid increase in enzyme activity and a severely sprouted kernel may be subject to several thousand times as much activity as a sound kernel. High enzyme activity can cause severe problems in baking (lower absorption, sticky doughs, and, in extreme cases, sticky crumb).

Of course the ultimate test of a wheat's suitability for breadmaking is to see if flour milled from it is capable of producing good bread. However, in selecting a suitable test baking procedure one must first make some difficult decisions: what end-product, what formula and what process to use for testing? For the flour miller these decisions may be relatively straight-forward since he wants his flour to do well using the formulae and processes used by his customers to make an acceptable end-product as judged by local standards.

In the case of an exporting country such as Canada, whose wheat is used in over sixty different countries, the challenge is to choose a series of baking tests that will give as much information as possible about performance over a wide range of baking conditions. At the Grain Research Laboratory (GRL) three basic baking methods are used [(sponge and dough,[1] Canadian Short Process[2] and Remixed Straight Dough[3])] to assess the baking quality of new and up-and-coming wheat cultivars and to survey ongoing work related to new harvest and cargo shipments. These tests provide detailed information on mixing requirements and performance using both lean and rich formulae and under conditions of both long and short fermentation. Special simulation studies related to specific products in specific markets may also be carried out in order to determine any subtle factors influencing the suitability of a particular class or grade of Canadian wheat for that particular end-use. Some recent examples have included studies on Brazilian pão frances[4] and Colombian alīnado bread.[5]

3.4 Methodology and approaches used to evaluate bread and bread-wheat quality

Specific factors that could or should be considered when evaluating flours for breadmaking potential may be subdivided into processing factors and product factors. The evaluation and measurement of some of these factors at the GRL are illustrated below.

3.4.1   Processing factors

3.4.1.1   Water absorption.
(i) Farinograph absorption.   This is the amount of water that can be added to a fixed weight of flour to produce a dough having a certain maximum consistency when mixed at a specified (standard) speed in the Brabender Farinograph (a torque measuring, recording dough mixer). It is widely used as a preliminary estimate of baking absorption.

Farinograph absorption depends largely on two factors: protein content and damaged starch level. Care must be taken in evaluating absolute data, since considerable variation in absorption can be obtained when the same flour is tested on different instruments. A lower absorption than predicted from flour protein and damaged starch values indicates inherently lower absorption capacity in either or both, which is undesirable.

(ii) Baking absorption.   Optimum baking absorption, which is normally judged by the handling properties of the dough at panning, is the maximum amount of water that may be used consistent with a high yield of bread per unit weight of flour, satisfactory dough handling properties at panning (i.e. no stickiness problems at the divider, rounder and moulder), and satisfactory bread quality. Baking absorption may or may not be closely related to Farinograph absorption for which only initial mixing is considered. Baking absorption is increased by certain ingredients such as milk powder and is influenced by softening that may occur during subsequent resting of the dough.

Baking absorption is less influenced by starch damage than is Farinograph absorption and is largely a function of gluten protein content.[6] High starch damage may even have a negative effect on baking absorption, particularly when sufficient α-amylase is present and the baking method involves a relatively long period of bulk fermentation.[7]

Some spring wheat varieties, notably some semi-dwarfs of Mexican parentage, have an inherently lower baking absorption than expected for the protein content, even at an equivalent starch damage level.

3.4.1.2   Mixing requirements for optimum dough development.
(i) Mixing time and energy.   Rheological tests using such instruments as the Brabender Farinograph and the Chopin Alveograph are normally carried out on unyeasted doughs. More pertinent to the baking process itself is information on mixing requirements and dough-handling properties during breadmaking.

Mixing curves can be obtained by monitoring electrical power or torque during the dough mixing stage of baking. Power or torque may be integrated with time to give energy values. Weak flours (as opposed to strong flours) normally require less intense (slower speed) mixing to achieve adequate dough development and a peak dough consistency is reached with shorter mixing time and less total energy.[8] Some overly strong flours may require a more intense mixing action than is provided by conventional dough mixers.

The advent of short baking processes in the USA (Amflow and Baker) and the UK (Chorleywood) in the 1960s required that the GRL become knowledgeable and experienced with these new baking systems as they relate to Canadian wheat. A significant part of our research programme involved the study of dough mixers and the phenomenon of dough development. Over a 10-year span, instruments and machines needed for these studies were designed and constructed. Research centred around factors affecting the development of dough structure during bread making. Some basic concepts related to mixing requirements for optimum bread quality were elaborated and provided insight to study further the dough formation process. The effects of mixing intensity and work input were studied using the GRL pin mixer and a laboratory-scale programmed dough-mixing unit, designed and built by the GRL.[9] This permitted examination of such factors as mixing intensity, temperature, pressure and work input on dough development. This work illustrated the two basic requirements of 'intensity' and 'work level' to achieve properly developed doughs.[8]

Some countries make use of the dough brake, which consists of a pair of motor driven steel rolls having a variable gap (spacing) to develop bread doughs. The dough is repeatedly fed through the gap of the sheeting rolls in order to work and stretch it. In studying the implications of the mechanical development of bread dough by means of sheeting rolls,[10,11] the GRL 1000 mixer[12] and a modified stand of sheeting rolls were used to compare directly the bread obtained from each system of development. Comparable bread was obtained with sheeting rolls using only about 15 to 25% of the work required by a mixer. It was concluded that much scope may exist for modification of mixer design to allow for the development of doughs in a more efficient manner.

Discovery and study of the phenomenon of 'unmixing'[13,14] triggered the postulation of the relationship between developed and undermixed dough and consolidated concepts elaborated from earlier work on dough develop- ment. 'Unmixing' is the apparent reverse of dough development brought about when a dough mixed to peak consistency at high speed is mixed for a further period of time well below the minimum speed required for optimum development. At the reduced speed the dough changes in character from a shiny elastic mass capable of being stretched into thin sheets to a rough, lumpy dough lacking in cohesiveness. Bread processed from such doughs resembles bread obtained from a severely undermixed dough. All mixers impart two types of action that tend to work against each other. One is the stirring action, which may be considered as destructive to the formation of the sheet-like structure of the dough. The other is the stretching and working of the dough which has to be of sufficient intensity to overcome the detrimental effects of the stirring action and promote the formation of a laminar sheet structure. Sheeting rolls have no stirring action to negate beneficial stretching and squeezing action and therefore there is little wasted energy. Further studies have shown that mixer action may have some beneficial effects in 'opening-up' the dough to atmospheric oxygen. By contrast, when doughs are developed by sheeting rolls there is no appreciable atmospheric oxygen incorporated with this action and any oxidation must come from added chemicals.

Several means of obtaining mixing curves from commercial mixers were devised and these were published[15,16] in addition to studies of processing of direct interest to commercial baking.[17,18]

(ii) Mixing tolerance. The shape of a mixing curve will normally provide a clue to mixing tolerance. Flours with a desirably short mixing requirement may have less tolerance to under- or overmixing. However, flours with a wide tolerance to overmixing may have undesirably long mixing requirements.

Thus the need for a flour of good 'all round' baking quality to produce acceptable bread over a wide range of processing conditions, otherwise known as tolerance becomes apparent. This implies primarily tolerance to mixing and fermentation conditions that are different from those considered optimum, but is much more easily discussed than measured.

It is well known that strength factors associated with physical or chemical flour and dough properties are interrelated with those strength factors associated with baking quality. GRL researchers have been particularly interested in the relationship between mixing requirements and baking quality.[19] Because strong flours generally have longer mixing requirements than weak flours, there is a tendency to equate long mixing requirements with good baking quality and even to assume that long mixing requirements are a necessary evil that must be suffered in the interests of bread quality.

For both mixing requirements and dough-handling properties it is a case of 'not too little, not too much.' Flours with short mixing requirements do not usually have good tolerance to overmixing or to fermentation, may do poorly in baking methods involving long fermentation and will probably produce doughs that are over extensible and soft.[20] By contrast, flours with very long mixing requirements may produce bucky doughs and may do poorly under conditions of slow speed or minimal mixing, particularly when this is combined with short fermentation time. Flours that do well over a wide range of baking conditions and produce doughs with well-balanced physical properties normally have mixing requirements that are neither unduly short nor unduly long.

3.4.1.3 Oxidation response and requirement. This question is particularly important when high volume bread is produced in short process methods. The mechanism of action of the oxidation and reduction reactions that are a key to proper dough development are extremely complex and, as yet, not fully understood. Suffice it to say that for any given baking process and flour there may be an optimum level and proportion of fast and slow acting oxidants for maximum loaf volume and best loaf appearance. Many millers and bakers are concerned about recent, current and impending moves to ban the use of potassium bromate in breadmaking.

Recent studies in this laboratory have been directed towards increasing our understanding of the effects of oxidants and other improvers at different stages of the baking process. Dough and loaf height trackers have been designed and constructed to study expansion of doughs during fermentation, proofing and baking.[1,21] Addition of ascorbic acid and sodium steroyl-2-lactylate (SSL) were required to optimize the quality of Brazilian hearth bread.[21] The improving action of these ingredients was attributed to the effect upon oven rise which occurred over a longer period of time than when these improvers were not used. Development of desirable break and shred characteristics is also associated with the improving action. Preliminary results indicate that four common oxidants (ascorbic acid, potassium iodate, azodicarbonamide and potassium bromate) all produce similar improving effects for pan bread. All oxidants increased the time over which dough expansion occurred in the oven.[22]

3.4.1.4 Dough handling properties. As indicated earlier, dough-handling properties at the final make-up stages of baking – dividing, rounding, moulding and panning – may range from soft and sticky for doughs from weak flours to extremely inelastic and tough for doughs produced from flours that are overly strong. Well-balanced dough properties that fall between these two extremes are required for baking.

At present, dough-handling properties must be determined subjectively by a trained laboratory baker. We have developed a dough sheeting and handling property indicator[23] that provides objective, non-destructive measurements of extensibility, resistance to extension, work required for sheeting and an index of the gas retention of doughs at time of sheeting in baking tests. This latter instrument has been modified to provide results automatically in the form of a paper tape read-out in direct reportable units.

3.4.2 Product factors

Size of a loaf may be measured by estimation of its volume by seed displacement in a simple volume measuring device. External appearance, internal crumb structure and crumb colour must all be estimated subjectively by trained laboratory bakers. Specially designed bread scoring tables with built-in top lighting are used which simulate northern daylight for crumb colour evaluation, and diffuse side lighting for scoring crumb structure. We are hopeful that instrumental, objective measurement of these factors will be developed in the near future. Crumb texture and softness/firmness can be measured objectively using a number of compression-type instruments such as the Instron Universal Tester. The GRL Compression Tester[24] has been developed to measure bread crumb properties such as deformation and resiliency.

Keeping quality is not normally taken into account when evaluating flour baking quality since, for any given flour and water absorption level, staling or crumb firming may be slowed by incorporation of emulsifiers and other ingredients. Bread flavour is also not a quality factor that is normally evaluated since bread flavour is heavily dependent on the formula and process used, and particularly on the oven baking stage and crust character (colour, crispness, etc.) Of course, freshness is an all important factor for the consumer. The attractive smell of freshly baked bread is one of the prime factors that caused supermarkets to introduce in-store bakeries.

3.5 Bread-wheat quality – from whose point of view?

Where does all this discussion lead? The question addressed finally here is how the different players with an interest in bread may define bread wheat quality?

The consumer wants a loaf of bread that suits his or her personal taste in terms of aroma, flavour, mouthfeel, crust and crumb characteristics, specific volume, etc. Choice of product will depend on whether the bread will be eaten fresh or over a period of several days, and whether it will be eaten plain or toasted, or used for sandwiches. Growing consumer preoccupation with healthy and nutritional eating habits and response to TV commercials and other advertising may dictate choice of brown, wholemeal, multigrain or some other variety of bread over white bread and may even lead to a willingness to pay a higher price for bread (e.g. made from organically-grown wheat).

Above all, the consumer is likely to want to obtain the same quality product every time he shops at his usual store. Previous taste experiences will lead him to prefer a particular brand or type from a particular bakery or supermarket. This in turn will lead to the expectation that every loaf he buys will be as good as the last one. In short, having made his choice, he wants no surprises. The choice of end-product may be very wide, but, if one ignores the major complication of price, consistency is probably the most important factor.

For the baker, also, the overriding concern is that every batch of flour of the same brand should perform in the same way as the previous one. He does not want to have to make major adjustments in his formula or processing conditions in response to variability in protein content, gluten strength or other factors. His definition of a flour with good breadmaking quality is one that will perform as expected over the range of processing conditions used and produce an acceptable loaf; again, with no surprises.

The miller probably faces more challenges than the baker with respect to uniformity and consistency. He must ensure that all batches of the same brand or type of flour meet rigid specifications and perform satisfactorily for his baker customers. The miller may have to cope with considerable variations in wheat quality. The quality of individual wheat lots can vary very widely due to differences in variety, growing location, growing and harvest conditions, damage caused during growing, harvest, or subsequent processing or storage (e.g. heat damage due to improper drying). If the miller is using a high proportion of locally grown indigenous wheat, he may have to test every batch to determine whether it should be accepted and, if so, how it should be segregated and stored. For imported wheat the miller will have some knowledge of the average quality he will receive and will rely heavily on uniformity and consistency of ongoing supplies; here again, no surprises.

Finally, the supplier (exporter) should put all these pieces of information together and conclude that the most important quality factor of all is uniformity and consistency. There may be many combinations of protein content, hardness and gluten strength that will suit specific end-uses, but for any given grade or segregate, it is of paramount importance that the quality of each shipment be as close to the average as possible.

Canada offers a wide choice of wheat quality to potential buyers through no less than eight classes of export wheat. The two soft classes (Canada eastern White Winter wheat and Canada Western Soft White Spring wheat) have relatively weak gluten and low protein content and are not thought of as bread wheats, even though a considerable proportion finds its way into flat bread production in the Middle East and Egypt.

Canada Western Amber Durum wheat is our premium semolina/pasta wheat, but some durum is used for bread production in Italy and other traditional durum wheat areas.

Canada Prairie Spring wheat is sub-divided into red and white. CPS wheat has a protein content of around 11.5%, medium gluten strength and semi-hard kernel characteristics, it is well-suited to production of French bread and Brazilian-type pão frances as well as noodles for which this high-yielding wheat was originally intended.

The most recently established class of Canadian wheat is Canada Western Extra Strong Red Spring wheat. This is a hard wheat with a protein content of around 12.5%. Its main feature is a very strong gluten which makes it desirable as a blending wheat capable of carrying weaker wheats in breadmaking.

Canada Western Red Winter wheat has excellent milling quality. With a protein content of a little less than 12% and medium to strong gluten, it produces excellent bread in terms of loaf volume and crumb structure. It does, however, yield flour with somewhat lower water absorption than our premium bread wheat, Canada Western Red Spring (CWRS) wheat.

CWRS wheat averages around 13.5% protein over the long term, but a sophisticated protein segregation system allows Canada to offer it at two or three levels of minimum guaranteed protein content; normally 12.5, 13.5 or 14.5%. It is this class that is designed to do well in any baking situation. Due to its large production, averaging over 20 million tonnes annually, a rigid variety registration system, a grading system that protects the top milling grades from damage caused by adverse growing and harvesting conditions, a bulk handling system that promotes mixing and blending within grades and, finally, a geography that causes grain exports to be funnelled west through the Pacific and east through Thunder Bay and beyond, Canada has earned a reputation for uniformity and consistency which is appreciated by its customers.

To summarize simply, high-quality bread wheat has the following properties:

1. a protein content consistent with the needs of the miller (usually at least 11.5% on a 13.5% moisture basis),
2. it is hard rather than soft so that target starch damage levels can be easily achieved by the miller,
3. a desirable balance of gluten strength, strong enough to produce bold loaves of high volume potential but not so strong as to give problems of long mixing requirement or bucky doughs,
4. potential for producing flours with water absorption consistent with the needs of the baker,
5. it is sound with no problems of excessive enzyme activity, and
6. it will produce good bread over a wide range of processing conditions.

References

1. Kilborn, R. H. and Preston, K. R. (1981) A dough height tracker and its potential application to the study of dough characteristics. *Cereal Chem.*, **58**, 198–201.
2. Kilborn, R. H. and Tipples, K. H. (1981) Canadian test baking procedures. II. GRL-Chorleywood method. *Cereal Foods World*, **26**, 628–630.
3. Kilborn, R. H. and Tipples, K. H. (1981) Canadian test baking procedures. I. GRL remix method and variations. *Cereal Foods World*, **26**, 624–628.
4. Dexter, J. E., Preston, K. R. and Kilborn, R. H. (1987) Milling and baking qualities of some Canadian wheat classes alone and in blends with Brazilian wheat under Brazilian processing conditions. *Can. Inst. Food Sci. Tech. J.*, **20**, 42–49.

5. Dexter, J. E., Kilborn, R. H. and Preston, K. R. (1989) The baking performance of Canadian bread wheat classes using a Colombian high-fat high-sugar short process. *Can. Inst. Food Sci. Tech. J.*, **22**, 364–371.
6. Tipples, K. H., Meredith, J. O. and Holas, J. (1978) Factors affecting farinograph and baking absorption. II. Relative influence of flour components. *Cereal Chem.*, **55**, 652–660.
7. Tipples, K. H. (1969) The relation of starch damage to the baking performance of flour. *Bakers Digest*, **43**, 28–32, 44.
8. Kilborn, R. H. and Tipples, K. H. (1972) Factors affecting mechanical dough development. I. Effect of mixing intensity and work input. *Cereal Chem.*, **49**, 34–47.
9. Kilborn, R. H. and Tipples, K. H. (1969) Improved small-scale laboratory mixing unit. *Cereal Sci. Today*, **14**, 302–305.
10. Kilborn, R. H. and Tipples, K. H. (1974) Implications of the mechanical development of bread dough by means of sheeting rolls. *Cereal Chem.*, **51**, 648–657.
11. Kilborn, R. H., Tweed, A. R. and Tipples, K. H. (1981) Dough development and baking studies using a pilot scale dough brake. *Bakers Digest*, **55**, 18–19, 22–24, 26, 28–31.
12. Kilborn, R. H. and Tipples, K. H. (1974) The GRL-1000 laboratory dough mixer. *Cereal Chem.*, **51**, 500–508.
13. Tipples, K. H. and Kilborn, R. H. (1975) 'Unmixing' – the disorientation of developed bread doughs by slow speed mixing. *Cereal Chem.*, **52**, 248–262.
14. Tipples, K. H. and Kilborn, R. H. (1977) Factors affecting mechanical dough development. V. Influence of rest period on mixing and 'unmixing' characteristics of dough. *Cereal Chem.*, **54**, (1977), 92–109.
15. Kilborn, R. H. and Preston, K. R. (1981) Device senses changes in dough consistency during dough mixing. I. With Tweedy mixer. *Bakers J.*, **41** (1981), 16–19.
16. Kilborn, R. H. and Tipples, K. H. (1981) Device senses changes in dough consistency during dough mixing. II. With horizontal bar mixer. *Bakers J.*, **41**, 40–42.
17. Kilborn, R. H. and Tipples, K. H. (1979) The effect of oxidation and intermediate proof on work requirements for optimum short-process bread. *Cereal Chem.*, **56**, 407–412.
18. Kilborn, R. H. and Tipples, K. H. (1970) Studies with a laboratory-scale 'acceletron' heat and steam accelerator. *Bakers Digest*, **44**, 50–52.
19. Tipples, K. H., Preston, K. R. and Kilborn, R. H. (1982) Implications of the term 'strength' as related to wheat and flour quality. *Bakers Digest* **57**, 16–18, 20.
20. Tipples, K. H. (1979) The baking test: Fact or fiction. *Cereal Foods World*, **24**, 15–18.
21. Kilborn, R. H., Preston, K. R. and Kubota, H. (1990) Description and application of an experimental heat sink oven equipped with a loaf height tracker for the measurement of dough expansion during baking. *Cereal Chem.*, **67**, 443–447.
22. Yamada, Y. and Preston, K. R. (1992) Effects of individual oxidants on oven rise and bread properties of Canadian short process bread. *J. Cereal. Sci.*, **15**, 237–251.
23. Kilborn, R. H. and Preston, K. R. (1982) A dough sheeting and molding property indicator. *Cereal Chem.*, **59**, 171–174.
24. Kilborn, R. H., Tipples, K. H. and Preston, K. R. (1983) Grain Research Laboratory compression tester: its description and application to measurement of bread-crumb properties. *Cereal Chem.*, **60**, 134–138.

4 Classification and grading

A. A. MACDONALD

4.1 Introduction

Quality control standards are the mainstay of Canada's reputation as an exporter of high-quality grain products. Because of the small population and relatively low domestic demand, the Canadian grain industry has always been geared to overseas markets and quality has provided a selling edge. Government administered regulations and standards for quality were well established before the Canadian Grain Commission evolved to harmonize the interests of grain farmers, handlers, exporters and importers.

Canada's grading and classification system is a distinctive feature of the overall approach to quality control. This presentation gives an overview of that system, with emphasis on wheat. It also highlights two other keys to quality control: varietal registration and cleanliness standards.

4.2 Why grade grain?

First, the purpose of grading will be considered. Why do most exporting countries grade grain and specifically, what are the objectives of Canada's system?

In any country, grain production is at the mercy of climate, with all its extremes and unpredictabilities. Even modern technology cannot compensate for uncontrollable growing conditions. These inevitably create regional variations in crop quality and yield within any given season and from year to year. A grading system permits the collection of grain of like kind and quality to facilitate both marketing and handling. Grading reinforces each link in the chain from farmer to agent, agent to processor and processor to end-user. In every transaction along the way, it serves as a value determinant.

To illustrate, consider grading as a prediction of how closely to optimum a homogenous lot of wheat will perform in the milling and baking processes. A baker succeeds when his customers are satisfied. Suppose a customer buys the same kind of doughnut from this baker time after time. In effect, he is 'grading' it highly enough to meet his taste and pays a given price. The baker must then produce a uniform product on a consistent basis to keep meeting the customer's expectations, charging his costs, labor and profit accordingly.

Continued success for the baker depends on access to a consistent and dependable supply of the same kind of flour he used to produce the original batch of doughnuts. The miller now has a potential customer for a certain grade of his flour, so he needs an ongoing and dependable source of wheat that meets his milling specifications.

Subject to many complex production and marketing conditions, a wheat farmer strives to produce crops that will be in demand and sell at a profit. Suppose he learns that doughnuts are selling well and decides to grow the class and variety of wheat suited to that market. Planting, growing and harvesting conditions prove ideal and his crop is free from insect damage and disease. The farmer's success now rests upon someone buying his wheat at an agreed price.

The farmer and buyer need a mechanism to help measure the value of the wheat value against the miller's requirements. The miller must in turn satisfy the baker, whose doughnuts will stand the ultimate test of consumer acceptance. We can thus define grading as the segregation of grain into parcels of defined quality to help determine price.

Two systems of grain grading are in use throughout the world: the fair average quality system and the numerical system.

The numerical system is most common. It separates grain into divisions of quality defined by grading factors. Each division is identified by a grade name or number and grain is bought and sold on the basis of these grades. Buyers select the specific quality they desire by grade name and number. In some grading systems such as the one Canada uses, importers accept a certificate of grade and no additional samples are required as evidence of quality. In other systems, customers demand additional safeguards such as a sample and analysis of the actual grain being shipped.

4.3   Canadian grading system

Like those in other exporting countries, including the United States, the Canadian grading system is federally legislated. The Canada Grain Act carries quality assurance even further, in that elevators and grain dealers must be licensed by the Canadian Grain Commission. As licensees, they must follow a set of regulated procedures in maintaining quality.

All classes of Canadian grain are separated by grade, and inferior quality is excluded from the export marketplace through the varietal registration system. This will be discussed in more depth later. Uniform and consistent quality is guaranteed within each shipment and from cargo to cargo, year after year.

The Canadian grading system facilitates both grain handling and marketing in several ways. By relating price to quality, grades simplify trading and help provide farmers with fair prices for their grain. They enable customers to obtain the same quality on a consistent basis and provide sufficient quality divisions to permit buyers to choose according to their needs. At the same time, Canadian grades effectively limit the number of quality divisions to accommodate the grain handling system.

4.4 Grade definitions and standards

Tables of grade definitions are the basis of assigning grades to grain. There are approximately 150 statutory grades applied to grain grown in Canada.

The Canadian grading system is based on visual assessments of grain samples. Standard samples are prepared as visual aids for trained grain inspectors. These standard samples visually interpret the grade definitions and reflect the growing conditions of the year for which they are prepared. They are selected annually by standards committees composed of farmers, exporters, processors, scientists and technical specialists. Separate committees establish standards for grain produced in Eastern Canada and Western Canada, respectively.

Two sets of standards are prepared. Primary standards are used at the producer level for almost all grades of grain. They illustrate the minimum acceptable levels of quality for grain delivered to primary and terminal elevators. Grain of a single grade moving in bulk through the handling system to export position now resembles a reasonable statistical average of the grade. Export standard samples are established at a higher level of quality than the minimum primary standard.

4.5 Grading factors

As mentioned, most major grain growing areas are subject to various conditions and extremes of weather during each growing and harvesting season. Canada is certainly no exception. The effects of degrading factors on end-use quality are impossible to determine visually and define precisely in our grade schedules. The Commission's Inspection Division relies heavily on scientific support from scientists in the Grain Research Laboratory. Different aspects of the work of the laboratory on wheat are reported in other chapters.

Although the grading system described here applies to wheat, rye, corn and barley, the next part of this overview focuses on wheat. Grading factors for wheat illustrate well the complexity of Canada's grain grading system. Five key grading factors are considered:

1 *Test weight* usually indicates grain plumpness and under normal conditions is not a limiting factor. Test weight, or grain density, is expressed

in kilograms per hectolitre for both domestic and export grades. It is determined by specific procedures using approved equipment. For example, No. 1 Canada Western Red Spring wheat has a minimum test weight requirement of 75 kg hl^{-1}.

2 *Varietal purity* is determined by the percentage of unregistered and deregistered varieties of the same class within the grade. By restricting varieties of inferior quality, varietal purity standards ensure that a class of wheat maintains its intrinsically high quality. Registered varieties not equal in quality to the varietal standards qualify only for the lower grades.

   Varietal purity does not necessarily mean that a sample must consist of only one variety. Mixtures of varieties are permitted, provided each variety in the mixture is equal in quality to the varietal standard. Currently, the varietal standard for the top four grades of red spring wheat is 'Neepawa', and for amber durum, it is 'Hercules'.

3 *Vitreousness* in hard red spring, amber durum and hard red winter wheat is defined as the glassy or shiny appearance that indicates hardness. Generally, the more vitreous a kernel, the higher its protein content. By definition, non-vitreous kernels are those that are broken, damaged, severely bleached or show signs of starchiness.

4 *Soundness* refers to the extent of overall physical damage. In most circumstances, it is the single most important grading factor. Sound kernels are well developed, mature and physically undamaged. Damage factors include frost, immaturity, weathering during harvest, disease and unfavorable storage conditions.

5 *Maximum limits of foreign material* refers to material other than grain of the same class remaining in the sample after dockage removal. This may include other cereal grains, inseparable seeds, thistle heads and pieces of stems. Foreign material is considered a degrading factor.

Three other factors are also important in wheat grading:

6 *Dockage* is the material that must be removed from a sample of grain before assigning an official grade. This material is readily removable using approved methods and equipment.

7 *Moisture content* is defined as the percentage of moisture in cleaned grain. Since the moisture content of each grain class is a grade determinant, all moisture-testing equipment must be operated strictly according to instructions and routinely checked for accuracy. The moisture content in trucklots of grain can be reduced by mixing with drier grain, aeration or artificial drying.

8 *Protein content* is the basis on which all No. 1 and No. 2 Canada Western Red Spring wheat is binned at export terminals using near-infrared analyzers. Final protein content is determined using the Kjeldahl process.

Protein segregation of hard red spring wheat has three objectives:

(i) To improve Canada's precision and flexibility in meeting world market demands for wheat of milling quality.
(ii) To identify the protein content of high-grade wheat as early in the marketing system as possible.
(iii) To reflect the market value of the protein of high-grade wheat directly back to the producer.

4.6 Grading consistency

It is extremely important that any sample receive the same grade at every inspection under every circumstance. Because grading depends upon visual examination and judgment of individual inspectors, various controls and procedures are imposed to ensure consistency of grading among inspectors.

4.6.1 Standardized grading procedures

Each sample of grain received by a terminal is identified and then prepared for grading using precision equipment approved by the Commission to remove dockage. A portion of the cleaned sample is tested for moisture using a Model 919 Moisture Meter. The sample is then examined for all grading factors. Grade is determined and recorded with notations to indicate reasons for assigning a specific grade. An official certificate of grade is issued for all samples obtained officially. An advice of grade is issued for unofficial submitted samples. Besides grade, these documents indicate moisture content as required. All export shipments are accompanied by the Certificate Final showing the quantity and grade of the shipment.

4.6.2 Selecting and training inspectors

Again, because the grading system is primarily visual, the work of the grain inspectors is intensive. Continuous staff training programs are necessary for all inspectors to achieve and maintain the required high level of skills and judgment.

Inspectors are recruited on the basis of aptitude and education. While serving a probationary period as assistants to inspectors, the trainees prepare samples for grading, analyze samples, conduct moisture tests and perform related work. After 11 months of training and successful completion of examinations, the trainees become qualified assistants to the grain inspector eligible to compete for inspector positions.

Newly qualified inspectors grade grain under supervision for up to two years. Then they must again pass theoretical and practical grading examinations to become fully qualified grain inspectors with a certificate recognizing their status under the Canada Grain Act. Extensive staff training programs are held regularly and inspectors take annual proficiency tests on varietal identification and other aspects of grading and inspection.

4.6.3   Sampling methods

It is very important that the sample taken for grading is representative of the complete parcel of grain, because the grade will be assigned on the basis of that sample. Grain officially inspected on entry to, or discharge from, a terminal elevator is officially sampled by an automatic sampler installed and operated under the supervision of the Inspection Division. The automatic sampler is a mechanical device that takes a representative sample from a stream of grain moving on a belt or through a spout. Where grain is bagged in warehouse lots, a special sampling device called a bag trier is used to sample grain at different points from a minimum of 20% of the consigned lot.

4.6.4   Standard equipment

Regularly maintained standard equipment is also essential to ensure uniform and consistent quality at the inspection offices of the Commission across Canada. Mechanical sampling devices used for grain shipped to or from elevators must be of a type approved by the Inspection Division of the Commission. Grain samples are cleaned and prepared for grading using equipment prescribed by the Canada Grain Regulations enacted under the Canada Grain Act. The Inspection Division specifies methods and procedures and routinely checks all equipment used for sampling and grading for accuracy and operational efficiency.

4.6.5   Sanitation

An extensive sanitation program helps maintain grain quality throughout the handling system. For example, many quality control procedures ensure that grain is shipped free from unacceptable levels of toxic residue.

It is illegal for a producer to deliver contaminated grain to a licensed elevator. If an inspector discovers evidence of chemical contamination in a car being unloaded at the terminal, he immediately orders that grain to be isolated. It is then tested by the Grain Research Laboratory and the owner of the grain is advised of disposal requirements. All vessel shipments are also tested for toxic residues.

The Inspection Division maintains entomology laboratories at each regional office in Canada. Cleanliness facilities at principal ports are continuously monitored by inspection staff. All rail car unload samples and composite samples of shipments from elevators are carefully examined. Inspection personnel periodically inspect terminal and transfer elevators and may order remedial action if necessary. All vessel shipments are tested for the presence of insects. If an insect problem is discovered, shipments from the source elevator are stopped until the infestation is eradicated.

4.7 Classification

The reader should now have a better understanding of grading, so that classification, again in the context of wheat, may now be explained. Classification recognizes the inherent differences in wheat classes. Examples are Canada Western Red Spring, Canada Western Amber Durum, Canada Prairie Spring, Canada Western Extra Strong Red Spring, etc. These classes all have genetic differences that make them more or less suitable for particular end-uses. For example, low-protein white wheats have high flour yields. They are ideal for the production of crackers but unsuitable for North American-style breads.

4.7.1 Varietal registration

This leads naturally to a discussion of varietal registration, the undisputed cornerstone of Canada's quality reputation. The technology and expertise needed to develop and evaluate new grain varieties and classes have evolved alongside Canada's grading system and the changing requirements of producers and consumers.

Canada is the only country with varietal quality standards named in its grading schedules. When the Canadian system is compared to others, varieties are a major topic. Varietal control is the number one factor influencing Canadian wheat quality.

Because varieties differ in their basic quality characteristics, it is not possible to determine visually whether grain that looks good also has good intrinsic quality. So Canadian plant breeders have had to build genetically into new varieties the qualities end-users and processors require. By allowing only approved varieties into the top grades, the desired quality parameters and uniform quality levels can be achieved by visual inspection.

Before registration is possible, a new Canadian grain cultivar must undergo years of testing for agronomic performance and resistance to disease. Once thousands of unacceptable lines have been discarded, the remaining few are scientifically evaluated and compared to preset targets. For every grain class there is a quality yardstick which all new varieties must meet or beat. Only these varieties which fall within a rather narrow range of quality characteristics are selected for registration and then only those that can be class-identified and segregated at the point of farm delivery will be supported for licensing.

In Canada there are between one and 19 varieties registered in each class of wheat with a very slow rate of new releases. By contrast, in the United States, there are well over 150 varieties of hard red winter wheat planted, and a number of new varieties released each year.

Canadian varieties are rigidly monitored on the basis of visually distinguishable characteristics. Because the grading and quality control system hinges on this requirement, it also poses a challenge to plant breeders in developing new varieties.

The shape and color of wheat or barley kernels within a given class must look different from those of all other classes. For example, while a particular new variety of red spring wheat may perform well in milling and baking tests, it must also show kernel characteristics unique to the wheat class red spring to qualify for registration. Canadian plant breeders have been extremely successful in meeting both quality and distinguishability objectives.

The main reason varieties must visually resemble others in their class is to guarantee and maintain purity in all grain shipments. The mixing of non-registered and registered varieties would not only devalue the export grain, but also Canada's reputation for consistent high quality.

Why is consistency so important? More and more processing industries around the world are modernizing their equipment and processes. Plant adjustments are costly enough without having to adapt further to changes in raw materials from shipment to shipment. For example, Japan is one of Canada's most discriminating customers. Japanese wheat processors regularly evaluate and compare end-use quality characteristics for material they import. Canada always obtains top marks for consistency. On the other hand, millers elsewhere have experienced serious problems with 'non-milling' varieties that were indistinguishable visually from varieties of acceptable quality.

4.8   Cleanliness standards

The consistently high level of Canadian wheat quality over the years is due largely to varietal standards. Canadian cleanliness standards are another very important quality control aspect.

As discussed earlier, practically all export grain must be physically cleaned at terminal elevators prior to shipment. The Canadian climate and rather short growing season combine greatly to minimize insect infestations and the need for chemical fumigation. Thus, Canadian grain is recognized as having particularly low pesticide residue levels. Similarly, Canada's short, hot, usually dry growing season limits the field infections of plant diseases which can potentially lead to toxic contamination.

4.9 Quality averaging

Canada's quota delivery and train-run system also allows delivery of a consistent product. It moves all grain from all areas gradually, thus ensuring that quality averaging takes place within each grade, with consistency between customers as well.

4.10 Objective versus subjective measurements

As this chapter has emphasized, Canada's wheat grading and classification system is based on visual distinguishability. Plant breeders have been extremely successful in genetically engineering intrinsic value while maintaining kernel class characteristics. Thus, numerical grades can be assigned according to a visual interpretation of environmental conditions.

Grade definitions enable grain inspectors to evaluate visually factors that cannot be precisely measured. For example, maturity and weather damage in cereal grains are subjective determinations made in comparison with standard or guide samples. Measurable factors, such as admixtures of foreign material and severe weathering damage, are assessed using standard equipment and procedures.

While visual grading remains practical, effective and efficient, the demand to build objective tests into grades increases. Rapid instrumental methods are a desired form of technical support to the visual grading system, although not intended to replace it. In the past decade, the search for faster, simpler and affordable test methods has been a priority shared by the Commission's Inspection Division and Grain Research Laboratory.

One important objective is to find a rapid, non-visual test which inspectors could use in grain terminals to screen for α-amylase and preharvest sprouting. Consumer demands are making this area of research more and more vital, with near-infrared techniques showing the most potential.

The need for rapid instrumental tests to help inspectors identify grain varieties is also urgent. While varietal identification will always hinge on some degree of visual distinction, plant breeders would have an easier job if the visual requirement were relaxed. Inspectors are finding it ever more difficult visually to distinguish cultivars by class. Weather conditions can affect kernel appearance so dramatically that varietal composition cannot be quickly determined. More pressure has come with the proliferation of non-registered wheat varieties in the early 1980s, and now, freer trade with the United States.

There is optimism that new objective measurement tools will soon enhance the work performed by Canadian grain inspectors. At the same time, the day when grading technology has advanced to the point where grain grown anywhere in the world can be evaluated in the same way by everyone in all systems is eagerly awaited.

With a continuing world market for high-quality Canadian grains, the Canadian Grain Commission is making every effort to stay abreast of end-use quality requirements in those increasingly sophisticated markets. In consultation with our customers and the industry, the commission will adapt the grading system as necessary to uphold Canada's reputation for grain quality into the 21st century.

5 Recent developments in flour milling

H. J. SCHOCH

5.1 Introduction

The growing environmental consciousness, safety and social requirements of staff, tightening legislation on human food, fierce competition, and rising labour and energy costs are some of the challenges the flour milling industry has to face today. The pressure on profit margins has led to mergers and the concentration of production on mills of higher capacities.

5.2 General requirements

The above business environment has instigated numerous new developments in milling engineering, aiming mainly at lower production costs, complete process automation, improved working conditions and ultimate product control.

5.2.1 Lower production costs

Lower production costs may be obtained through:

- design of high-capacity equipment to allow the construction of larger mill units,
- reduced investment by saving building volume and costs,
- design of equipment requiring less maintenance to save labour and to reduce the duration of production stoppages, and
- saving of energy by, for example, the application of air recycling systems for all screenroom equipment and automatic air controls for the mill pneumatic system.

5.2.2 Automation

Automation of the entire process aims to:

- improve management control,
- assure flour consistency by precise adjustment of equipment, quick production changeovers and repeatable milling results, and
- achieve optimization of flour quality,

Figure 5.1 Transflowtron wheat measures.

- provide quality and production records,
- eliminate unsociable working hours and overtime, and
- assist and reduce personnel.

5.2.3   Improved working conditions

Staff may enjoy improved working conditions with:

- the design of easy-to-adjust and safe operating equipment,
- noise reduction, and
- special regard to plant safety.

5.2.4   Product control

Ultimate product control may be obtained by:

- the application of on-line quality control systems to ensure consistently high performance of milling products in bakeries and to eliminate consequential damages, and
- implementing the best possible sanitation, through use of appropriate equipment and plant design.

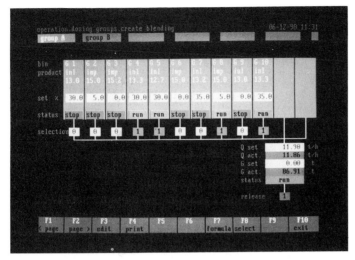

Figure 5.2 PC for wheat gristing and stock control.

5.3 Developments in equipment and processes

The length of this chapter does not allow us to deal in detail with all the achievements made during the past years. However, let us have a close look at those which had a significant impact on progress in the various phases of the technological process.

5.3.1 Screenroom

Because of growing environmental consciousness, farmers must use less pesticides and fertilizer. As a result, wheat of poorer quality is supplied to the mill. This requires an improved screenroom performance and well equipped wheat blending systems in order to obtain clean wheat grists with uniform properties; the basic prerequisites for a satisfactory mill performance.

The new Translowtron wheat measures (Figure 5.1) which work on the principle of loss-in-weight feeding, not only meter and accurately feed each wheat according to PC controlled blending recipes, but also supply data for stock and production control (Figure 5.2).

Large air volumes which were required for proper wheat cleaning consumed much energy and filter area. The development of an air/dust separator based on the use of centrifugal forces as well as the vector of dust velocity for dust extraction, has allowed re-use of the separation air. The consequent application of this principle to the aspirator (Figure 5.3), combinator (Figure 5.4) and dry stoner (Figure 5.5) has resulted in

Figure 5.3 Aspirators with air recycling system.

Figure 5.4 Combinator.

Figure 5.5 Dry-stoner with air recycling system.

substantial savings in energy consumption of up to 50% for a 22 tons hr^{-1} screenroom. This has been made possible by a drop in fresh air requirement of over 90%. In addition, a reduction in building space could be obtained since fewer filters and fans are needed.

5.3.2 Mill

The merger of milling companies has led to a concentration in production plants and the construction of flour mills with much higher capacities, for example of 400 to 600 tons per unit per day. This requires the development of equipment of corresponding size. The new square-type sifter (Figure 5.6) has a sifting area of up to 75 m^2, approximately 36% more than the largest type available before. Its application reduces the number of equipment units, saves space and lowers costs.

In past years, the conventional milling technique has been optimized to such a degree that the extraction rates achieved in the efficient new plants are difficult to improve upon. For this reason it comes as no surprise that news of dehulling processes with promising results have attracted a lot of attention.

Figure 5.6 Square-sifters each with sifting area up to 87 cm^2.

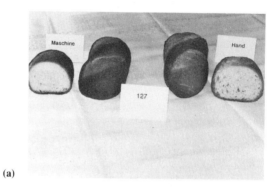

(a)

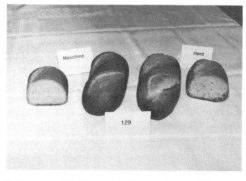

(b)

Figure 5.7 Baking test with flours. (a) Conventional milling. (b) Debranning.

A large number of our own tests involving dehulling of various wheat types to different degrees and with varying conditioning procedures show that (see Figure 5.7):

- the ash curves are identical or inferior to those of the conventional milling process and that the ash content of the first reduction flours is high,
- the extraction calculated on wheat prior to dehulling is the same or lower compared with the results in modern mills without dehulling, and
- the baking quality of flours with the same colour or ash is the same.

At the same time, the energy requirement and capital expenditure are substantially higher.

On the other hand, a new milling process with double grinding of stocks without sifting, which was introduced about three years ago, has proved to be a striking success (Figure 5.8). More than 20 mills were built and commissioned all over the world using this new technique, mainly for the 1st/2nd break, A/B and D/E streams. The milling results are the same as those with the conventional system except that:

- the power and air requirements of the pneumatic system are reduced by approximately 13% due to the elimination of the lifts for the 2nd break, B and E streams,
- the sifting area is reduced by 8%,
- the requirement in building volume is reduced by 15 to 20%, and
- the capital expenditures are lower.

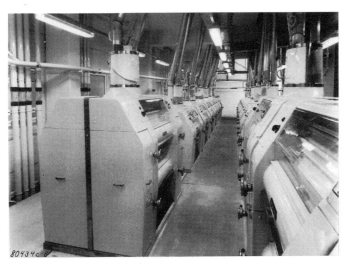

Figure 5.8 Eight roller mills.

Figure 5.9 Field proved NIR on-line instrument.

Encouraging field experience with this new system confirms the results of the thorough testwork made prior to its introduction into the mills. The experience also shows the potential of the new process for application to further mill passages.

5.3.3 Quality assurance

A baker remains a happy customer so long as the flour supplied to him has a consistent quality and is clean from any infestation. However, human food legislation enforces these demands and continuous on-line flour quality records can save the miller consequential damage claims.

The field proven near-infrared (NIR) instrument (Figure 5.9) continuously records the protein and moisture contents as well as the ash/colour value. In addition, closed loop control systems are available for the control of:

- the protein content by accurate addition of vital gluten, and
- the moisture content by addition of water with computer-controlled water feeders.

Figure 5.10 Efficient sterilator.

The treatment of flour and a sterilator (Figure 5.10) kills moths and weevils, reduces the risk of infestation in flour silos and ensures a long shelf life. The machine can be fitted into a pneumatic lift. Compared with the earlier models, it considerably reduces the pressure loss in the blowlines.

5.3.4 Flour silo

In many flour mills around the world, flours and additives are mixed by means of batch-blending systems permitting the production of tailor-made flours and the achievement of sophisticated production records. The disadvantages of such systems are their high power and building requirements. Because of the latter, the incorporation of such systems into existing plants can only be made with high capital expenditure.

New solutions are possible due to developments in electronic continuous-flow scales like:

- the differential dosing scale (Figure 5.11) (loss-in-weight feeder) for capacities from 2 to 3000 kg h^{-1}, and
- the Transflowtron/Transflowmeter (Figure 5.12), which are continuous flour-flow controllers (measurers of the loss-in-weight principle), feeding flow capacities from 1 to 40 t h^{-1}.

These new tools permit the construction of fully automatic, continuous blending plants.

Figure 5.13 shows an example of a possible flow chart for a flour blending system with Transflowtron blending control. The different flours are

Figure 5.11 Differential dosing scale (loss-in-weight feeder).

Figure 5.12 Transflowmeter, continuous flour-flow controllers.

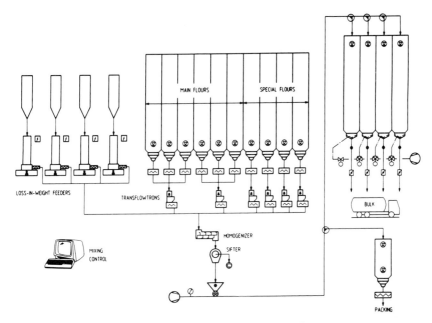

Figure 5.13 Flow chart of continuous flour blending system.

weighed and blended in continuous streams together with preselected quantities of additives. The batch mixer is not required and a self cleaning homogenizer or mixing screw complies with the homogenizing requirements. This system allows the precommissioning of packing or bulk loading jobs. A computer control system manages production without personnel, provides production and quality records as well as stock and loading statistics.

The major difference between batch-blending and continuous-blending systems is in the accuracy especially at the beginning of the cycle. With batch blending, even the first batch includes all components as each single one is accurately weighed and the batch thereafter thoroughly mixed. With continuous blending there is an inaccuracy owing to start-up lags and spouting at the very start of the job for approximately 250 kg. However, this problem can easily be solved by the use of a 'start-up' hopper, into which the flour is fed at the beginning of the job and is fed back to the stream afterwards. For this reason, the continuous blending system is most suitable for large plants in which quantities of at least 4 tons of each flour are blended.

5.4 Summary

Free trade agreements, quality management standards, environmental requirements and social demands of personnel are only some of the challenges facing the milling industry today. New equipment and techno-logical developments may not only help the flour millers to cope with these tasks but may increase profitability by efficiency, high product quality standards and reduction of production costs.

6 Recent research progress in bread baking technology

W. SEIBEL

6.1 Historical overview

It is believed that the Egyptians discovered bread baking.[1] Mural paintings show that sourdough was already used in Egypt as early as the 13th century BC. Findings from eastern Mediterranean regions suggest that bread baking was already known around 1800 BC. From Egypt, the art of breadmaking came to Israel and then to Greece. Around 776 BC the conclusion of the Olympic Games was celebrated with a feast that included bread made from wheat and barley. In Greece leavening of the bread was achieved with sourdough or soda; a special bread cult existed and Demeter, the goddess of bread, was revered.

After the Romans had conquered the Greeks, they learned from the Greeks how to make bread. Around 100 BC hundreds of small bakeries and even some baking companies were in the city of Rome. The white bread was preferred by the rich, whereas the common people consumed the dark bread and wholegrain meal bread.

Around the turn of the century, the Romans introduced bread to central Europe. Bread and cake were mainly prepared in monasteries. Brewer's yeast was developed in the 15th century and was also used for the production of bread. The production of baker's yeast began in Europe around 1900.

In Europe, research in the area of baking technology started in the second half of the 19th century and was intensified at the beginning of the 20th century. Until the middle of this century, the main objectives were the development of machinery adapted to the size of the company and the improvement of processing techniques. The suppliers of raw materials steadily improved their products and today, high-quality milled cereal products are available to bakeries worldwide.[2,3]

Baker's yeast with good fermentation properties is offered everywhere as pressed yeast or dried yeast. The baking-improver industry contributed significantly to the improvement of bread quality by developing new products for special purposes.

In the last decade, research in the area of baking technology has passed through various phases, depending on nutritional concerns, application of new or improved techniques, and consumer trends. This will be the subject of the following discussion.

6.2   Nutritional recommendations

For the last one or two decades, it has been known that a great discrepancy exists in the western industrial countries between recommended and actual intake of nutrients. In general, the consumption of fat, animal protein and sugar is too high. In contrast, our diet is low in dietary fiber, complex carbohydrates, and certain minerals and vitamins. These unhealthy nutrition habits led to a variety of diseases, the treatment of which is costly. Therefore, different countries have developed nutritional recommendations to compensate for the imbalance of nutrients in the diet and to improve the health of the public. With respect to baking technology, it is important that all nutritional recommendations include an increase in the consumption of cereal-based foods, mainly of wholegrain products. Recommended cereal based foods include:

- cereal foods such as whole grains, flakes, extruded products, breakfast cereals, granola bars, etc.,
- baked goods, especially bread and rolls from wholegrain, and
- pasta products, especially those made from wholegrain.

These nutritional recommendations have led, in the last decade, to an increase in the consumption of bread and other cereal-based products in a number of countries. The increase in per capita consumption of bread in Germany is shown in Figure 6.1.

Marketing studies have indicated that it is necessary to develop a wide variety of products with excellent sensory properties and good freshness in order to increase the consumption of cereal foods. Therefore, in many countries, studies have been carried out to incorporate additional ingredients into the classic formulae and thus, to increase the variety of the products. Optimum formulations were developed to produce cereal foods according to the nutritional recommendations.

6.2.1   Expanding the varieties of bread

Apart from bread varieties which are specific for each country,[4-9] new bread specialities have been developed during the last decade. This was achieved in different ways by:

- the use of grains other than wheat and rye,
- including materials of animal and plant origin in the formulation, application of old and new baking processes, and
- preparation of breads with different nutritive value, including dietetic and vitaminized breads.

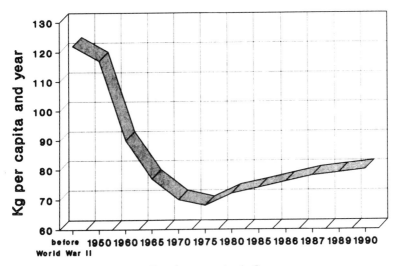

Figure 6.1 Bread consumption in Germany.

6.2.2 Use of non-bread grains

In many countries bread is only made from wheat. In some countries of central Europe rye is also used. Apart from these two bread grains, barley, oat, rice, sorghum, and triticale can be incorporated to a certain extent in the formulation. Bread formulations have been developed that, besides normal bread wheat (*Triticum aestivum*), also include German spelt wheat (*Triticum spelta*) and durum wheat (*Triticum durum*). The optimum form for applications of these non-bread grains (meal, grits, flour, etc.) has been examined.[10–12] It is also known that a pretreatment by swelling (cold or hot) can result in a sensory improvement of the bread. Table 6.1 summarizes the results of a comprehensive study, indicating optimum addition levels and recommended forms of non-bread grain during bread production. The breads produced are generally characterized by a somewhat different appearance and specific taste. In sensory cases, these different quality attributes may make bread more appealing to the consumer and thus promote its consumption.

Table 6.1 Optimum addition levels of non-bread grains during bread production

	20 parts	30 parts	40 parts
Oat flakes			x
Barley flakes	x		
Corn semolina	x		
Rice (cooked)		x	
Sorghum (cooked)	x		
Buckwheat grits		x	

6.2.2.1 Materials of plant and animal origin. Today, a wide variety of materials of plant and animal origin are available. Materials of plant origin include, among others, wheat germ, raisins, spices, brans, and especially, oil seeds.[6,7,11] For example, in German bakeries, more than 20 different kinds of oil seeds are used, with a minimum addition of 8 kg per 100 kg milled grains. The currently most important oil seeds are soy beans, linseed, sunflower seeds, and pumpkin kernels.

The most used materials of animal origin are dairy products such as full cream and skim-milk powder, butter, and sour-milk powder. Quark, whey, and yoghurt are also used. Butter is increasingly used in toast-bread production. It is added as normal butter, pure butter fat, or fractionated pure butter fats. It is always possible to replace fat by oils with better nutritional properties without encountering difficulties in processing. Addition of materials of animal origin imparts a special flavor to the bread, thereby contributing to an increase in bread consumption.

6.2.2.2 Special baking techniques. In the past, bread was baked in ovens heated with wood. Today, this procedure has been reintroduced as a result of trends motivated by nostalgic feelings among certain alternative consumer groups.

During the normal baking process, the loss of vitamins in the crust can rise to 30%. Therefore, special wholemeal breads are baked in steam ovens which prevent crust formation and hence the loss of vitamins. If a thick crust is desired, baking in a stone oven is recommended. In Germany, for some types of bread, a process called 'Gersteln' is used. Before the dough is put into the oven, it is treated for a short time over an open flame to create dark specks on its surface.

6.2.2.3 Breads with altered nutritive value. In order to meet the requirements of specific nutritional recommendations, baking technology can be modified by increasing or decreasing certain nutrients or components of the bread formula. However, special claims can only be made when the nutritive value of the new product is significantly different from the nutritive value of a regular bread.[13] With respect to protein enrichment, the protein content should be at least 20% of dry matter.[6,7] This is about twice as much as the protein content of a regular bread. The increase can be achieved by addition of ingredients high in protein, such as a mixture of vital gluten, milk, and soy protein.[14,15]

With respect to carbohydrate reduction, a decrease in the carbohydrate content of 30% is required in order to justify the claim. This reduction can only be obtained with a significant increase in protein.[6] More important than a carbohydrate reduction is an energy reduction. In recent years, a 'light' wave has started worldwide and various sectors of the food industry are offering 'light' products.[19] The term 'light' can mean different things.

Whereas in the past, 'light' was related in Europe to the digestibility of foods, today the consumer is using the term with regard to calorie/energy-reduced foods, mainly to a low-fat content. In Germany, the energy reduction of foods is regulated by law. In the area of cereal foods the energy content of an energy-reduced bread should be less than 840 kJ or 200 kcal per 100 g bread. This reduction in energy can be accomplished by replacement of white flour by wholegrain flour, by increasing the dietary fiber content through an addition of high-fiber fractions (see later) or by increasing the moisture content of bread by substances with high water-binding capacity, e.g. hydrocolloids. Wholegrain breads are particularly suited for energy reduction. If the formula contains considerable amounts of fat (e.g. toast bread), fat can be replaced with suitable substitutes.

There has been a growing interest in dietary fiber-enriched breads since the health-minded consumer has become aware of calorie reduction and other beneficial effects of fiber ingredients in baked products. However, in the western industrially developed countries there is still an imbalance between the recommended fiber intake of 30 g per day and the actual intake of 20 g per day. To eliminate this deficit, a part of the daily bread consumption should consist of the wholegrain flour/meal breads (7–8% (as is) dietary fiber) instead of white breads (3–4% (as is) dietary fiber).[16–18] Another possibility is the production of high fiber breads by the addition of high fiber fractions with dietary fiber content between 50 and 90% (dry matter). When these fractions are incorporated in the formula at levels between 5 and 10%, a significant increase in the total dietary fiber content can be achieved. Currently, high fiber fractions from different sources are available such as cereal brans, legume brans, fruits, potatoes, and beets. It should be mentioned that during the last few years, much research has been carried out on enzyme-resistant starch, which appears to have beneficial effects similar to those of dietary fiber.[18]

Recently, breads with a relatively high proportion of soluble fiber have been prepared. High soluble-fiber sources are, for example, oat and barley bran and sugarbeet fiber. A bread high in soluble fiber might contribute to lowering the cholesterol level, thereby reducing the risk of cardiovascular disease. However, caution is still advised in evaluating the efficiency of dietary fiber in the treatment/prevention of this and many other diseases.

Considerable progress has been made in the area of dietetic breads.[6,7] Hydrocolloids such as low protein starches, carob bean gum, and/or swelling starches are used in the production of these breads. For celiac patients, gluten-free breads were developed. These products are based on rice, sorghum, buckwheat, corn, and wheat starch. The formula should contain no ingredients from wheat, rye, triticale, barley, or oats.

Among the dietetic breads, vitaminized breads play a minor role. These breads are usually prepared from flours of low extraction rate. By adding mixtures of vitamins or products naturally rich in vitamins such as wheat germ or vitamin yeast, the level of vitamins is increased to that of a wholegrain flour bread.

Increase in salt intake has been related to an increase in blood pressure. In response to a recommended reduction in salt intake, bread formulae with half of the normal sodium chloride level, i.e. maximum of 250 mg per 100 g bread, have been developed without affecting sensory properties of the bread. It is even possible to produce breads with strongly reduced sodium content, lower than 40 mg sodium per 100 g bread. This requires the elimination of salt from the formula and might be accompanied by processing problems such as impaired dough properties, poor gas retention capacity, and bland taste. Wholegrain meal and wholegrain flour breads are especially suited for production of sodium-reduced breads because these types of breads are characterized by an intensive and aromatic taste. Also the rye-containing sourdough breads are well suited for a sodium chloride reduction.

6.2.2.4 *'Bio' breads (alternative breads).* Worldwide there is a tendency, especially among young consumers, towards healthier eating patterns.[20] This trend implies the production of bread without additives and the consumer would expect that the cereal raw materials are produced without the use of fertilizers, pesticides, etc. The European Community has passed a law that regulates the growing conditions of plants used for the alternative 'bio' foods as well as the composition of 'bio' breads.[21] Normally, not all bread formula ingredients are produced and are available to comply with these regulations. Therefore it is permitted to incorporate up to 5% of conventionally produced ingredients in the formulation. Thus, an alternative 'bio' wholewheat bread can be prepared by using wholemeal wheat ('bio'), sodium chloride (non 'bio'), yeast (non 'bio'), and water. No chemical additives are used. A 'bio' bread is more expensive than a regular bread and some unscrupulous bakers may be tempted to sell a regular bread as a 'bio' bread at a higher price. To protect the consumer, strict regulations are necessary to control the production and distribution of alternative 'bio' foods.

6.2.2.5 *Optimization of bread formulations.* In the past, it was difficult for baking technology to respond to specific nutritional recommendations with an appropriate variety of baked goods. It was always necessary to carry out numerous experiments with different recipes to find an optimum formulation. Today, the application of response surface methodology (RSM) has made it possible to develop easily the optimum formulations for different purposes.[22,23] For example, we focused on the production of wholewheat bread with high loaf volume, good crumb elasticity, and improved shelf life. The ingredients included wholewheat flour, sodium chloride, yeast, and certain additives such as emulsifiers and hydrocolloids.

Figure 6.2 Optimum wholewheat bread formulation. MDG = monodiglyceride, DAWE = diacetyltartaric ester of a monoglyceride, CMC = carboxymethylcellulose, VV = control.

Figure 6.2 shows the result of the RSM study. A wholewheat bread with optimum volume and crumb properties could be produced according to the calculated formulation that included, among others, 0.3% monodiglyceride, 0.6% diacetyltartaric ester of monoglyceride, 0.15% guar gum, and 0.5% carboxymethylcellulose.

Application of RSM was especially advantageous regarding economical considerations. The traditional way of testing different formulae requires about 100 experiments whereas only 40 experiments are needed according to the RSM. This systematic approach can be applied in many areas of baking technology.

6.3   Processing techniques

As in the past, bread is still prepared by mixing, fermentation, and baking. However, improvement in processing techniques and equipment has facilitated the production and led to an improvement in product quality.

6.3.1.   Milling technique

Baking technology has benefited from progress in milling techniques. As a result of nutritional recommendations aimed at increasing the consumption of dark bread, there has been a growing demand among consumers for a well-leavened wholegrain bread. In Europe, wholegrain breads were traditionally produced from coarse milled products (wholegrain meal). These breads were somewhat dense because of the low gas-retention capacity of the wholegrain meal dough. Wholegrain meals were normally produced with a hammer mill or a stone mill. To prepare well-leavened wholemeal wheat bread from these milled products a wheat with a soft kernel structure was recommended. The suitability of kernel structure can

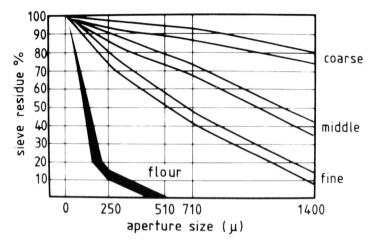

Figure 6.3 Particle size distribution curves of milled wheat products. The terms 'fine', 'middle' and 'coarse' refer to the wholewheat meal.

be easily determined by measuring the test weight of the respective wholemeal. The lower the test weight of the wholemeal, the better the wholemeal is for breadmaking and the better the leavening of the bread. However, an optimum leavening of a wholegrain bread can only be achieved by the use of wholegrain flours. Figure 6.3 illustrates the particle size distribution curves of wholegrain meals and a wholegrain flour which are typical for Germany. Wholegrain flours can only be prepared by roller milling. During this process, which is similar to the milling of white flours, the endosperm is carefully milled to a flour containing small endosperm particles. Breads prepared from wholewheat products milled from the same wheats but differing in particle size are shown in Figure 6.4. The bread made

Figure 6.4 Wholewheat bread. From left to right: coarse-, middle-, fine-wholewheat meal, wholewheat flour.

from wholewheat flour is much more leavened and has a higher volume than breads made from wholewheat meals of different particle sizes.

6.3.2 Developments in dough and breadmaking techniques

During the last few decades, attempts have been made to shorten the breadmaking process by developing suitable processes and equipment. Continuous dough mixers, the application of high-temperature fermentation, and baking with the combined use of different energy sources (e.g. convection heat, high frequency, microwaves, etc. are typical examples). The experience with these developments, however, shows that the bread quality, especially the aroma and taste (flavor), has not been improved at all. Today we know that certain minimum production times are necessary to bring out the full aromatic taste. Prestages such as predoughs and sourdoughs as well as optimum baking times can be used to improve the flavor of bread.

Since the freshness of bread is one of the most important quality criteria in terms of consumer acceptance, higher water additions are used worldwide. This results in more sensitive doughs which require special equipment. Therefore, development in equipment has been directed to machines that process the dough very carefully, mainly during the dividing and moulding stage.

6.3.3 Computer-assisted baking machinery

With the introduction of the electronic processing of data in bakeries, it is now possible to control automatically the entire production process, from the combination of formula ingredients to the specific conditions during baking. The TNO (The Netherlands Organization for Applied Scientific Research) mixing system[24] was designed to determine optimum mixing time by measuring the increase in temperature. Possibilities also exist to control automatically the baking program in the oven by following certain parameters. Some progress was also made in automating the quality control work. It is possible to determine the degree of browning of baked products with sensors. However to ensure the quality of the products, a sensory evaluation performed by trained panelists is still required.[7]

6.3.4 Extrusion cooking

Today baked goods are not only prepared in baking ovens. With the technology of HTST (high-temperature short-time) extrusion cooking, baked products, especially flat breads, can also be made.[25]

Extrusion cooking is a versatile process and it has a wide range of applications in the production of cereal-based foods. The potential of

Table 6.2 Nutritional value of flat bread

	Protein digestibility (%)	Net protein utilization	Biological value
Whole rye meal	68.9	40.7	59.1
Flat bread, baked	71.7	34.8	48.5
Flat bread, extruded	70.4	38.6	54.9

extrusion cooking lies in various areas, e.g. simplifying and modifying the manufacturing of existing cereal products and developing new ones. The extruder is unique in working with relatively dry materials. The cereal mass moves through the machine and receives a short input of thermal and mechanical energy. This results in changes in structure and composition, similar to conventional food-processing operations such as baking. Thus, extrusion cooking can be used for production of existing foods where the shape and form of the product can be reproduced at the outlet die and where the energy transfer is able to impart sensory attributes to the end-product similar to the baking process.

When the extrusion technology is used for production of flat bread, the specific volume of the bread is doubled compared to conventional flat bread and the mouthfeel is different. A great variety of extruded flat breads is now available on the market. Some of the breads are manufactured with very lean formulae based on wholegrain flours or white flour, others are produced with very rich formulae including different protein sources and various milk products. In Germany, about 10–15% of the flat breads on the market are processed by extrusion cooking.

There are, of course, differences in the nutritive value between conventional and extruded flat bread due to the shorter heating period in the extruder. The differences in nutritive values are shown in Table 6.2.[26]

6.3.5 Freezing techniques

Freezing techniques have become increasingly important in the area of baking technology. In the past mainly the end-products, i.e. bread and rolls, were frozen, whereas today freezing has shifted to the dough stage. The freezing technique helps to solve problems with regard to freshness because fresh baked products, mainly rolls, can be offered during the entire opening hours of the bakery.

The minimum shelf life of the crispy German standard roll is 4–6 h. Hence, it is possible to bake several times a day. In the normal bakery the following production schedule has been established. Early in the morning, doughs for rolls are prepared to be baked immediately. Then, after the kneading process, dough pieces are cooled to approximately 4° C, stored

Figure 6.5 The effect of dough freezing on German rolls. From left to right: control (without dough freezing), without prefermentation and with final fermentation after thawing the dough, with prefermentation, and final fermentation after thawing.

and baked later; other doughs are formed into dough pieces, prefermented and then frozen and stored at −18° C. In the frozen state, the dough pieces can be transported to the shops for production of crispy rolls. Much progress has been made in determining optimum roll formulation and freezing conditions. It is now possible to place a frozen dough piece directly in the oven at 200–220°C and to bake a roll within 22 min. The results are shown in Figure 6.5.[27]

6.3.6 Sourdough procedure

For the production of rye breads an acidification of the dough is necessary because of biochemical reactions that require a lower pH. This can be achieved by the use of a sourdough. Apart from this traditional use, sourdough has also been used in recent years for production of baked goods from wheat such as the French baguette and the Italian pannetone. In the past, each bakery prepared its own sourdough which was carried through several stages. This led to the development of one-, two-, or three-stage sourdoughs.

Nowadays, defined sourdough starters are available for automated sourdough equipment.[28] A recent development is a continuous sourdough manufactured according to a patented method. However, this process has only been used successfully in some large companies.

6.4 Labeling of bread

Over the last decade, the consumer has become increasingly informed about bread. Within the European Community, labeling of bread is standardized. Labeling includes the type of bread, price, weight, ingredients, and shelf life. Although nutritional labeling is not required, it is often provided.[13] If nutrition information is included on bread labels, the protein, fat, carbohydrate, and energy (kJ or kcal) contents as well as the total dietary fiber content should be indicated. Wholegrain meal breads have a dietary fiber content of 7–9% (as is), white breads of 3–4% (as is).

Information on the presence of contaminants is usually not given. It is known that the amount of heavy metals and chemical residues (e.g. pesticides) in milled cereal grain products is low and that a further reduction occurs during breadmaking due to the dilution effect of added water.[29] However, precautions should be taken not to bring contaminants into the bread by other ingredients, e.g. cadmium in sunflower seeds.

6.5   Quality assurance standards – sensory evaluation

In the future, European market quality assurance standards will take effect to comply with EC standards 29000 to 29004.[30] Within this quality control system, sensory evaluation of raw materials and bread for the purpose of quality assessment will play a decisive role. Fortunately, significant progress has been made during the last decade in the sensory evaluation of foods with emphasis on bread.[7]

Generally, an international 5-score system is used for the sensory evaluation of bread. Compared to a good quality standard bread, differences and impairments in quality are determined and described. The individual quality criteria are the following:

- form, overall appearance,
- crust, surface,
- leavening/crumb grain,
- elasticity,
- structure, and
- smell and taste/flavor.

The panelists participating in a sensory evaluation are well trained.[31] Therefore, the evaluation is as accurate and reproducible as chemical or physical methods. In Germany, it is possible to obtain a certificate which, based on a successfully completed examination, qualifies the person to evaluate organoleptically bread, soft wheat products (e.g. cake and cookies), and other cereal-based foods such as flakes and breakfast cereals.

6.6   Convenience products

The variety of convenience products used in the bakery has increased continuously due to high labor costs and the lack of qualified employees. Convenience products are offered in the sector of flour mixtures and concentrates, cake mixes, doughs, and baking ingredients. The formulae of these convenience products are, in general, optimized. Thus, the final products have a good or excellent quality. The mixtures comply with food law regulations and the food producer obtains information about labeling and marketing/advertisement of the foods made from convenience products.

6.7 Summary

Bread is still baked from fermented dough. During the last decade bread recipes have become increasingly adjusted to dietary guidelines in different countries. Other grains besides wheat and rye are now used as raw material. Ingredients of animal origin are also found in modern recipes.

There is a distinct trend worldwide towards decreasing the amount of additives. There has been remarkable progress in process engineering, leading to computerized bread production. Modern baking machines do not damage the dough; machine cleaning is increasingly important.

Bread freshness is the essential point in marketing. Extrusion cooking in production of crisp bread is being used. There is an increasing market in frozen doughs. Bread continues to be a major component in dietary recommendations.

References

1. Seibel, W. and Brummer, J.-M. (1991) Historical development of sourdough applications in the Federal Republic of Germany. *Cereal Foods World*, **36**, 299–304.
2. Drews, E. and Seibel, W. (1976) Bread baking in other countries around the world. In *Rye: Production, Chemistry and Technology*, (ed. W. Bushuk) AACC, St. Paul, Minnesota, pp. 127–178.
3. Seibel, W. (1978) Developments in baking technology. In *Cereals – 78: Better Nutrition for the World Millions*, (ed. Y. Pomeranz) AACC, St. Paul, Minnesota, pp. 187–201.
4. Miller, B. S. (1980) *Variety Breads in the United States*. AACC, St. Paul, Minnesota.
5. Seibel, W. (1987) Possibilities and limits for standard baking tests. In *Cereal in the European Context*, (ed. E. D. Morton) VCH, Weinheim/New York, pp. 315–322.
6. Seibel, W., Menger, A., Brummer, J.-M. and Ludewig, H.-G. (1985) *Brot und Feine Backwaren*, DLG-Verlag, Frankfurt a.M.
7. Deutsche Landwirtschafts-Gesellschaft (1991) *DLG-Prufbestimmungen fur Brot und Feine Backwaren*, DLG-Verlag, Frankfurt.
8. Seibel, W., Brummer, J.-M. and Stephan, H. (1978) West-German Bread. In *Advances in Cereal Science and Technology. Vol. II*, (ed. Y. Pomeranz) AACC, St. Paul, Minnesota, pp. 415–453.
9. Greenwood, C. T., Guinet, R. and Seibel, W. (1981) Soft wheat uses in Europe. In *Soft Wheat: Production Breeding, Milling and Uses* (ed. W. Yamazaki and C. T. Greenwood) AACC, St. Paul, Minnesota, pp. 209–266.
10. Seibel, W., Brummer, J.-M. and Neumann, H. (1991) Herstellung von Haferbroten. *Getreide Mehl u. Brot*, **45**, 140–145.
11. Brummer, J.-M. and Seibel, W. (1991) Spezialbrote mit Nicht-brotgetreidearten und/oder Olsamen. *Getriede Mehl u. Brot*, **45**, 45–50.

12. Brummer, J.-M., Morgenstern, G. and Neumann, H. (1988) Herstellung von Hafer-, Gerste-, Mais-, Reis, Hirse- und Buchweizenbrot. *Getreide Mehl u. Brot*, **42**, 153–158.
13. Verordnung uber Nahrwertangaben bei Lebensmitteln (Nahrwert-Kennzeichnungs-verordnung) vom 9.12.1977 (*BGB1 I*, **S. 2569**) zuletzt geandert am 30.5.1988 (*BGB1 I*, **S. 667**).
14. Pomeranz, Y. (1983) Protein enriched breads in USA. *Getreide Mehl u. Brot*, **37**, 277–281.
15. Jussef, M. M. and Bushuk, W. (1986) Breadmaking properties of composite flours of wheat and faba bean protein preparations. *Cereal Chem.*, **63**, 357–361.
16. Shogren, M. D., Pomeranz, Y. and Finney, K. F. (1981) Counteracting the deleterious effects of fibre in bread making. *Cereal Chem.*, **58**, 142–144.
17. Seibel, W. and Brummer, J.-M. (1991) Ballaststoffe und Backverhalten. *Getreide Mehl u. Brot*, **45**, 212–216.
18. Sievert, D. and Pomeranz, Y. Enzyme-resistant starch. (1989) I. Characterization and evaluation by enzymatic, thermoanalytic, and microscopic methods. *Cereal Chem.*, **66**, 342–347.
19. Seibel, W. (1991) Kennzeichnung bzw. Kenntlichmachung von Light-Produkten in der Zukunft. *Deutsche Backerzeitung*, **78**, 1049–1052.
20. Seibel, W. (ed.) (1990) *Bio-Lebensmittel aus Getreide*, Behr's Verlag, Hamburg.
21. European Community (ed.), EC-Regulation 2092/91; June 24, 1991.
22. Mettler, E. (1990) Experimentelle Studien der Emulgator- und Hydrocolloidwirkung zur Optimierung der funktionellen Eigenschaften von Weizenbroten. *Dissertation*, Universitat Gießen.
23. Mettler, E., Seibel, W., Brummer, J.-M. and Pfeilsticker, K. (1991) Experimentell Studien der Emulgator- und Hydrokolloidwirkung zur Optimierung der funktionellen Eigenschaften von Weizenbroten. 1. Mitt.: Rezepturauswahl – Einfluß der Emulgatoren und Hydrokolloide auf das rheologische Verhalten von Weizenmehlteigen. *Getreide Mehl. u. Brot*, **45**, 145–152.
24. Sluimer, P. (1989) Automatisierung des Knetprozesses fur den handwerklichen Backbetrieb. *Getreide Mehl u. Brot*, **43**, 24–26.
25. Meuser, F. and Wiedmann, W. (1989) Extrusion plant design. In *Extrusion cooking*, (ed. C. Mercier, P. Linko and J. M. Harper) AACC, St. Paul, Minnesota, pp. 91–155.
26. Harmuth-Hoene, H. E., Seibel, W. and Seiler, K. (1986) Die Proteinqualitat von Knackebrot und extrudiertem Trockenflachbrot aus Roggenvollkornschrot. *Zeitschrift fur Ernahrungswissenschaft*, **25**, 196–204.
27. Seibel, W. (1991) Tiefgekuhlte Weizenteige in der handwerklichen Backerei. *Backerzeitung*, **23**, 19–21.
28. Meuser, F., Faber, Ch. and Mar, A. (1989) Kontinuierliche Fuhrung von Sauerteigen mit verschiedenen Mikroorganismenkulturen und unterschiedlichen Roggenmahlerzeugnissen in einem zweistufigen Pilotfermenter. *Getreide Mehl u. Brot*, **43**, 328–337.
29. Ocker, H. D. and Bruggemann, J. (1991) Zur Bewertung der Schadstoffsituation des Brotgetreides. *Getreide Mehl u. Brot*, **45**, 6–10.
30. CEN (ed.) *Quality Management and Quality Assurance Standards*. EN 29000–29004/ISO 9000–9004, (ed. CEN) Rue Brederode 2, B-1000 Brussel, 1987.
31. Seibel, W. (1991) Erfahrungen und Ergebnisse des 18. DLG-Sensorik-Seminars. *Getreide Mehl. u. Brot*, **45**, 250–252.

7   Wheat proteins: structure and functionality in milling and breadmaking

J. D. SCHOFIELD

7.1   Introduction

Wheat flour proteins have long been known to be crucial in relation to breadmaking quality, both protein quantity and quality being important.[1-3] The major wheat endosperm storage proteins, the gluten proteins, comprising two prolamin groups, gliadin and glutenin, have been studied intensively because they confer the viscoelasticity on doughs considered essential for breadmaking quality. Qualitative differences in their composition and properties account for much of the variation in breadmaking quality between wheat cultivars. Glutenin, comprising polymers with subunits linked by disulphide bonds, is particularly important.

A great deal of research has now been done to characterise gluten proteins and to determine the molecular basis for their functional properties and for the variation in those properties amongst wheat cultivars, and to determine the genetic control over that variation. Enormous advances in our knowledge of these proteins have been made over the last 10–15 years in particular, with the application of modern methods of analysis and a higher level of research effort.

We now have a reasonably good picture of the basic structural features of the different types of polypeptide that comprise the gluten protein fraction.[4-7] The genetics of these polypeptides is also reasonably well known,[8-10] and our knowledge of the effects of individual polypeptides on breadmaking quality is increasing rapidly.[3,8-10] Despite the achievements of recent years, much essential detailed knowledge remains to be gathered about the polypeptides of the gluten complex.

Much less is known of the polymeric structure of the subfraction of gluten proteins known as glutenin. Progress in this direction has inevitably had to await a detailed description of the individual polypeptides that comprise this fraction. There is also a great deal still to be learned about the interactions between the different gluten proteins themselves and also their interactions with flour components other than proteins and how these affect breadmaking quality. The effects of dough processing upon flour protein structural properties and their relation to functional behaviour are still only known in relatively vague terms at the molecular level.

Besides breadmaking quality, the flour milling performance of wheat is also of considerable technological importance.[11] Much less is known about the factors that affect intercultivar variation in milling quality than those that affect breadmaking quality but starch damage level, which in turn affects water absorption and dough rheology is an important characteristic. A minor protein fraction associated with the starch granule surface has been implicated recently as having a role in controlling grain endosperm texture (hardness), an important component of milling quality.[12] As a result, research is now under way in a number of laboratories to characterise this protein fraction.

The present status of our knowledge regarding the protein components that play important roles in flour milling and breadmaking quality is summarised here.

7.2   The importance of gluten proteins to breadmaking quality

Early research from Finney's laboratory[13,14] showed that breadmaking quality was strongly influenced by the level of flour protein, the higher the protein content, the better the breadmaking quality. This work also indicated that qualitative differences exist among the proteins of different wheat cultivars such that at any given protein level the breadmaking qualities of some cultivars were always poorer than those of others. Across the range of protein contents normally found in wheat flours, breadmaking quality (as indicated quantitatively by loaf volume) increased linearly with flour protein content for all cultivars. The gradients of the regression lines (called 'protein responses') differed amongst cultivars, however, such that poorer cultivars had lower protein response values than better cultivars. Similar observations have been made by others, and, in Figure 7.1, these effects of protein level on the breadmaking quality of two Canadian cultivars are shown clearly in work from Bushuk's laboratory.[15]

Flour can be separated into several major fractions, i.e. starch, gluten and a water-solubles fraction, under mild aqueous conditions, and the freeze-dried fractions subsequently recombined such that the breadmaking performance of the reconstituted flour is similar to, and in some cases the same as, that of the original unfractionated flour.[13,16–18] Using this technique, and interchanging the fractions from flours of different cultivars, it has been shown that, while the starch and water-solubles fractions are necessary to produce bread of normal quality in an optimised breadmaking test, they do not contribute significantly to variation in breadmaking quality among wheat cultivars.[19] Rather, the variation is accounted for by qualitative differences in the gluten protein fraction.[13–17]

More recent work has shown that the properties of other flour components, such as the binding of polar lipids as assessed by extractability from

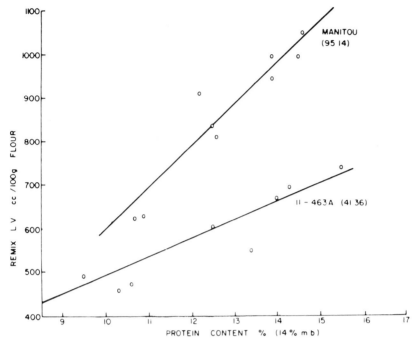

Figure 7.1 Effects of flour protein content on the breadmaking qualities (Remix Loaf Volume – LV) of good (Manitou) and poor (II-463A) Canadian wheat cultivars. Reproduced with permission from Bushuk, ref. 15.

flour with apolar solvents, e.g. *n*-hexane, may also be related to the intercultivar variation in breadmaking quality.[20–23] Nevertheless, this contribution is generally acknowledged to be smaller than that of the gluten proteins.

As a result of the demonstrations that there are major functional differences in the gluten protein fraction, much research has been done to determine the nature of gluten proteins and the molecular basis for these differences. An important application (probably the most important medium term application) of such knowledge is in breeding wheat cultivars with improved breadmaking quality, and the work on chemical characterisation of gluten proteins has been complemented by studies to define their genetical control and recently their molecular biology. Progress in these areas has been impressive.

7.3   Classification of gluten proteins

Estimates of the proportion of wheat flour proteins accounted for by the gluten protein fraction, which represents the main storage protein reserve of the wheat seed, range from about 50% to about 80%.[24,25] In general, flours with higher protein contents have higher proportions of gluten proteins.[26,27] Gluten may be prepared from flour by forming a dough and then gently washing away the starch as a suspension (often referred to as a 'starch milk') in an aqueous phase that contains water-extractable albumin and salt-extractable globulin proteins. During this washing out procedure, the gluten protein fraction forms a cohesive viscoelastic mass with properties somewhat akin to chewing gum. Although protein enriched, gluten prepared in this way is by no means pure. The protein content is generally not more than about 75–80%, the bulk of the remainder being accounted for by residual starch, lipid and hemicellulose (pentosan).

Gluten has been considered as comprising two major groups of proteins, present in approximately equal proportions, which, in fact, are quantitatively the two most important groups of proteins in flour itself. These are the gliadin fraction, which is extractable in concentrated aqueous alcohol solutions, commonly 70%(v/v) ethanol, and the glutenin fraction, extractable after gliadin extraction with dilute acid, most often dilute acetic acid, or dilute alkali.[6,25,27,28] Some protein usually remains unextracted after alcohol and acid extraction, and is referred to as 'residue protein'.[29] Its amount varies according to the extraction procedures used. Under standardised extraction conditions, its amount also varies from one wheat cultivar to another.[29,30] It may contain variable proportions of unextracted glutenin.[28]

Traditionally, the gliadin fraction has been considered as the prolamin fraction of wheat protein, defined on the basis of its extractability in aqueous alcohol solution and its very high contents of proline and glutamine.[25,27,28] Glutenin has been considered as belonging to a different protein fraction, the glutelin fraction, defined by its insolubility in alcohol solution and by its extractability in dilute acid or alkali.[25,27] The terms prolamin and glutelin are generic terms applicable to similarly extracted protein fractions from all cereals, whereas the terms gliadin and glutenin describe those two groups specifically in wheat.

We now know that the lower molecular weight wheat glutenin polymers have appreciable extractability in aqueous alcohol solutions,[31–33] and that the alcohol extractability of glutenin is greatly increased when reducing agents and dilute acid are included in the aqueous alcohol solution.[25] The amino acid sequences of the constituent polypeptides of glutenin are also closely related to those in the gliadin fraction.[6] Furthermore, many of the polypeptides in the glutenin and gliadin fractions are closely related genetically in that they are encoded by genes at the same chromosomal loci,[8–10] and both are deposited within protein bodies in the endosperm.[34,35] In view of this, it is beyond doubt that the polypeptides of the gliadin and glutenin fractions are very closely related and that both, in fact, belong to the prolamin group, as suggested some time ago by Miflin *et al.* and Shewry *et al.*[34,36] The difference in alcohol extractability between gliadin and

Wheat Gluten Proteins

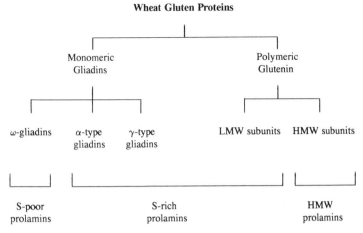

Figure 7.2 Classification of wheat gluten proteins based on structural homologies and genetical relationships. After Shewry and Miflin, ref. 4.

glutenin is due to the fact that the former occurs as monomeric proteins, whereas the latter largely comprises high molecular weight polymeric proteins, whose component polypeptides are linked by interchain disulphide bonds.

On the basis of the structural and genetical relationships among the different polypeptides in the gliadin and glutenin fractions, Shewry's and Miflin's group[4,6,37] has proposed a novel scheme for classifying gluten proteins (Figure 7.2). All are regarded as prolamins, and they are subdivided into three groups, (a) sulphur-rich prolamins, corresponding in older classification schemes to α- and β-gliadins (referred to collectively in Figure 7.2 as α-type gliadins), γ-gliadins and the low M_r (or low molecular weight (LMW)) subunits of glutenin, (b) sulphur-poor prolamins, corresponding to ω-gliadins, and (c) high molecular weight (HMW) prolamins, corresponding to the high M_r (or high molecular weight (HMW)) subunits of glutenin.

There is little doubt about the biological validity of this classification. Nevertheless, in relation to the functional properties of the different prolamin species, there is considerable merit in the argument for retaining the terms gliadin and glutenin to differentiate between the monomeric and polymeric forms, whose component polypeptides, although structurally related, appear to be largely distinct functionally.[2,38]

7.4 The nature of gliadin and glutenin polypeptides

Both gliadin and glutenin are complex and heterogeneous protein fractions, as indicated by two-dimensional electrophoresis[39] (Figure 7.3). Early

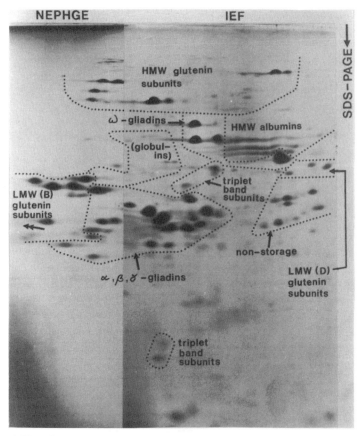

Figure 7.3 Two-dimensional electrophoretic map of wheat endosperm proteins from the cultivar Chinese Spring illustrating the complexity and heterogeneity of the polypeptide components of gluten and other proteins. Separation in the first dimension was by a combination of non-equilibrium pH gradient electrophoresis (NEPHGE, left hand side) and isoelectric focusing (IEF, right hand side) and in the second by discontinuous polyacrylamide gel electrophoresis in the presence of sodium dodecyl sulphate (SDS-PAGE). HMW = high M_r, LMW = low M_r. Reproduced with permission from Payne *et al.*, ref. 39.

attempts to fractionate them by the classical Tiselius moving boundary electrophoresis technique resulted in four components, named α-, β-, γ- and ω-gliadins being differentiated.[40] Later separations, using starch gel electrophoresis[41,42] and, more recently, polyacrylamide gel electrophoresis at acid pH,[43] gave greater resolution, and far more components were identified. It was possible to relate the polypeptides separated by the higher resolution techniques to the four groups separated by moving boundary electrophoresis, and gliadins are still commonly referred to as belonging to the α-, β-, γ- or ω-gliadin groups. By polyacrylamide gel electrophoresis in the presence of sodium dodecyl sulphate (SDS-PAGE), the relative molecular masses (M_rs) of the α-, β- and γ-gliadins fall within the range 30 000–40 000.[2,6] The α- and β-gliadins generally have lower M_rs than the γ-gliadins.[6] The α- and β-gliadins are also very similar to each other structurally, but exhibit some differences from the γ-gliadins (see below). For these reasons, the α- and β-gliadins are referred to together as α-type gliadins and the γ-gliadins as γ-type gliadins.[6] Both types contain appreciable levels of sulphur amino acids (2–3 mol %), and also possess intramolecular disulphide bonds. The ω-gliadins, in contrast, have higher M_rs of 44 000–80 000.[44–46] They are deficient, and, in some cases, completely lacking, in sulphur amino acids, but are rich in glutamine/glutamic acid (40–50 mol %), proline (29–30 mol %) and phenylalanine (8–10 mol %).[6] They are also the least charged of the gliadins. The α-type and γ-type gliadins have lower levels of glutamine/glutamic acid (34–42 mol %), proline (< 20 mol %) and phenylalanine (ca. 5 mol %).[6]

By some two-dimensional electrophoresis techniques, approaching 50 different gliadin polypeptides have been identified in a single wheat cultivar,[47] but allelic variation occurs among the gliadin polypeptides in different wheat cultivars, increasing the complexity of this group of proteins still further. This allelic variation provides the basis for cultivar identification procedures,[48] which are important commercially in quality control in both seed and grain trading.[49] It is also important scientifically in relation to determining the genetical relationships between the different gliadin species and relationships with functional quality.

Glutenin occurs as polymers, the individual polypeptides of which (usually called subunits) are linked by interchain disulphide bonds. Although not as heterogeneous as gliadin, in terms of the number of polypeptide types present, up to 19 different components are present, nevertheless, in any one cultivar.[50] As with gliadin polypeptides, allelic variation occurs in glutenin subunits between cultivars, and this is of considerable importance in relation to breadmaking quality variation in wheat.[8–10] The subunits are readily classified into two groups on the basis of differences in M_r (Figure 7.4). The high M_r subunits have M_rs estimated at between 80 000 and 160 000 by SDS-PAGE, depending on the particular subunit and the SDS-PAGE system used,[51] although the true M_rs determined by sedimentation equilibrium and amino acid sequencing are very much lower, from about 63 000 to 88 000.[52,53] They account for around 10–20% of the total prolamin protein and only around 5–15% of total flour protein.[7,54,55] Yet, apparently, they have a major effect on breadmaking quality or dough properties, and they have been studied extensively for this reason.

The low M_r subunits are present in much larger proportions than the high M_r subunits.[7,55] Their M_rs are similar to, although slightly higher than, those of α- and γ-type gliadins, as determined by SDS-PAGE, as are their amino acid compositions.[37] However, two-dimensional electrophoretic techniques[50] and amino acid sequence analyses[6,56] indicate that the low M_r

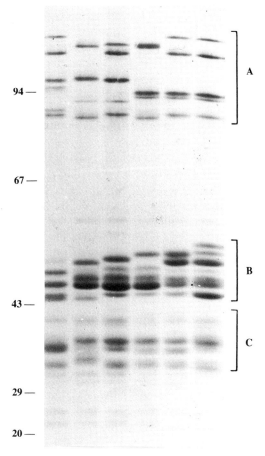

Figure 7.4 Fractionation of high M_r and low M_r subunits of glutenin of the wheat cultivars Halberd (lane 1), Chinese Spring (lane 2), Sunstar (lane 3), Gabo (lane 4), Hartog (lane 5) and Suneca (lane 6) by a one-step, one-dimensional SDS-PAGE procedure. The positions of M_r markers ($\times 10^{-3}$) are shown at the left side. The high M_r (or A) subunits are well separated from the low M_r subunits, and the latter can be classified into two groups, B and C subunits, the B group being of somewhat higher M_r. Reproduced with permission from Gupta and MacRitchie, ref. 115.

subunits are distinct from α- and γ-type gliadins. The high M_r subunits are significantly different from the low M_r subunits in terms of their amino acid compositions, containing higher levels of glycine and tyrosine and generally lower levels of sulphur amino acids, valine, isoleucine, leucine and phenylalanine.[6]

Most of the glutamic acid and aspartic acid residues in glutenin and gliadin polypeptides are present as the amide forms glutamine and asparagine.[57] This, together with the low contents of basic amino acid residues, means that

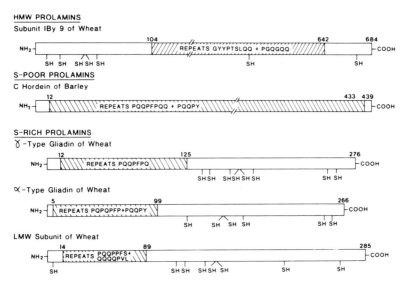

Figure 7.5 Schematic comparison of the structural features of glutenin and gliadin polypeptides, illustrating the presence in each of a central domain containing repeated amino acid sequences and of unrepetitive N- and C-terminal domains. ω-Gliadins are represented here by the structurally similar C-hordein of barley. Reproduced with permission from Tatham *et al.*, ref. 6.

these polypeptides have relatively low ionic character (i.e. they have relatively low charges) compared with other proteins.[3] Indeed, the ω-gliadins have been described as being amongst the least charged proteins known.[57] The gluten polypeptides as a group are also relatively rich in amino acids with hydrophobic side chains, implying a high potential for interaction via the hydrophobic effect. The presence of such high levels of glutamine and asparagine also implies a potential for extensive interaction via hydrogen bonds, since the amide groups of the side chains of those amino acids can act simultaneously as electron donors and acceptors.[1]

The amino acid sequences of the different types of glutenin and gliadin polypeptides have some general similarities (Figure 7.5). Each has a central domain or region that contains repetitive sequences flanked by unrepetitive N-terminal and C-terminal domains.[6] The lengths of these domains vary from one type of polypeptide to another, and, in some cases, the N- or C-terminal domains are very short, comprising only a few amino acids, whereas in others they may account for the bulk of the polypeptide. Also, the repeat motifs in the central domains are based on different polypeptide sequences in the different polypeptide types. These repeat motifs are clearly related to each other for the α- and γ-type gliadins, the low M_r subunits of glutenin (i.e. the S-rich prolamins of wheat) and the ω-gliadins (the S-poor prolamins).[6] The tetrapeptide motif PQQP occurs in all the S-rich prola-mins, but, in the α-type gliadins it is interspersed with a heptapeptide motif, PQPQPFP, and in the low M_r subunits with another heptapeptide, QQQQPVL.[6] The sequences of the S-poor ω-gliadin repetitive domains do not appear to have been elucidated fully, but can be inferred from that of the closely related C-hordein of barley. The main repeat motif here is the octapeptide PQQPFPQQ, although, at the N-terminal end of the domain, some related pentapeptides (consensus sequence PQQPY) occur.[6]

The unrepetitive C-terminal domains of the α- and γ-type gliadins and the low M_r subunits are also closely related, and most of the cysteine residues are conserved in these domains.[6] Of the S-rich prolamins, only the low M_r subunits contain a cysteine residue in the unrepetitive N-terminal domain, which could be related to the ability of these polypeptides to form polymers. In the S-rich prolamins the C-terminal domains are large and comprise just over one-half to about two-thirds of the polypeptides, whereas the N-terminal domains are only 5–14 amino acids long.[6]

All the major high M_r subunits of glutenin, of which there are either 3, 4 or 5 in any particular cultivar, have been sequenced or their amino acid sequences have been inferred from the base sequences of cloned genomic DNA. Two types of high M_r subunit are recognised, so-called x- and y-types (Figure 7.5), the former being somewhat larger than the latter.[58] The x-type subunits have a repetitive central domain between about 650 and 700 amino acids long.[5-7] They have nonapeptide, hexapeptide and tripeptide repeat motifs with consensus sequences of GYYPTSPQQ, PGQGQQ and GQQ.[5,6] In the y-type subunits the repetitive central do-mains are shorter (435 to about 540 amino acids). The tripeptide repeat motif is absent, the hexapeptide motif is the same as in x-type subunits, but the nonapeptide consensus sequence is GYYPTSLQQ.[5,6]

The repeat motifs in the high M_r subunits are unrelated to those in gliadin polypeptides and low M_r subunits of glutenin, but the sequences in the unrepetitive N- and C-terminal domains are homologous with those in gliadin and low M_r subunits.[5,6] The x-type subunits have N-terminal do-mains of 81 to 89 amino acids and C-terminal domains of 42 amino acids. The N-terminal domains of the y-type subunits are larger (104 amino acids), but the C-terminal domains are the same size as those of the x-type subunits.[5-7]

The cysteine residues in the high M_r subunits are located mainly in the N- and C-terminal domains, although several of the y-type subunits have a single cysteine residue towards the C-terminal end of the central repetitive domain, and x-type subunit 5 has one close to the N-terminal end of its central domain.[5-7] The N-terminal unrepetitive domain is richest in cys-teine residues. In x-type subunits there are three residues in this domain, and in y-type subunits five. The C-terminal non-repetitive domains of both x- and y-type subunits contain only one cysteine residue.[5-7] The N- and

C-terminal domains also tend to be richer in charged residues than the central domains, but poorer in glutamine and glycine.

7.5   Conformational structures of gliadins and glutenin subunits

A variety of physicochemical (e.g. spectroscopic and viscometric) and predictive techniques has been used to determine the conformational structures of gliadin and glutenin polypeptides.[6] The α-type gliadins appear to have compact globular-type structures, whereas the γ-type have more extended structures. The ω-gliadins and high M_r subunits of glutenin are even more asymmetrical, and form rod-shaped structures.[6]

The α- and γ-type gliadins and low M_r subunits of glutenin are similar in terms of their secondary structures. In the unrepetitive C-terminal domain there is a mixture of secondary structures. The studies suggest that substantial contents of α-helix and β-sheet are present, as well as random coil, but that some β-turns are also to be found.[6,59–64] The predominant conformation in the repetitive domains is the β-turn, however. In the α-type gliadins and low M_r subunits of glutenin, the β-turns are thought to be irregularly distributed, but in γ-type gliadins the repeat amino acid sequence motif that gives rise to the β-turn structure is highly conserved. In the γ-type gliadin species, the β-turns may occur with such regularity that a loose spiral conformation may form that may account for the more extended structures of these gliadins.[6]

The loose spiral structure may be of much greater significance in the S-poor ω-gliadins and high M_r subunits of glutenin, in which the central repetitive domain comprises the bulk of these polypeptides. α-Helix and β-structure are largely absent from ω-gliadins, but β-turns are abundant.[6,60,63] In the high M_r subunits of glutenin, the unrepetitive N- and C-terminal domains are rich in α-helical structure, but the central domains are rich in β-turns like the ω-gliadins.[5–7,61,65,66] Field et al.[66] have suggested that the β-turns in the central domains may be repeated to such an extent and be so regular as to result in the formation of a loose spiral structure similar to the β-spiral formed by a synthetic polypeptide based on the pentapeptide repeat motif of the elastomeric protein elastin present at high levels in some mammalian connective tissues with elastic properties (e.g. artery walls and ligaments). This spiral has a pitch of 0.945 nm and has 13.5 amino acid residues per turn. The hydrodynamic properties of high M_r subunits of glutenin are consistent with this proposed structure.[66]

Field et al.[5–7,66] have proposed a model (Figure 7.6) for the structure of high M_r subunits in which the repetitive central domain is largely in the form of a loose spiral based on repetitive β-turns, and this is flanked by globular, unrepetitive N- and C-terminal domains, which are α-helix rich and which contain most of the cysteine residues. They propose that the β-spiral

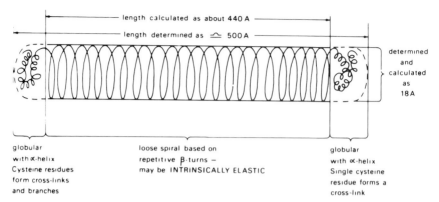

Figure 7.6 A structural model for the high M_r subunits of glutenin based on a β-spiral conformation for the central repetitive domain and α-helix rich globular N- and C-terminal domains. The β-spiral structure of the central repetitive domain may be intrinsically elastic. Reproduced with permission from Field *et al.*, ref. 66.

structure may be intrinsically elastic. Direct experimental evidence for the presence of the β-spiral structure in high M_r subunits has recently been provided thorugh scanning tunnelling electron microscopy.[67] The results of this work indicate that high M_r subunits do indeed have a rod-like structure with a diagonal periodic banding pattern that may correspond to the turns in the spiral. The dimensions of the observed structures are similar to those expected for a loose spiral based on β-turns.

7.6 The polymeric structure of glutenin

Glutenin exists as disulphide-linked polymers in its native state, and this polymeric structure has long been considered important functionally. Thus, redox reagents that may affect this polymeric structure have substantial effects on functional quality, including those such as potassium bromate, azodicarbonamide, ascorbic acid (after conversion to its dehydro form), cysteine and sodium metabisulphite, that have been used as ingredients in baked products.

Whereas the structures of the individual polypeptides (subunits) comprising glutenin polymers are now known in some detail, the polymeric structure of glutenin is little understood, although several models have been proposed. Over 20 years ago, Ewart proposed his 'linear glutenin hypothesis', which envisaged that glutenin comprised 'concatenations' of polypeptides that were linked head-to-tail by disulphide bonds in random order.[68-70] An important feature of Ewart's model was that the individual subunits of glutenin were envisaged as having a conformation that could be stretched when a shearing force was applied, but that would recoil elastically when the

(a)

(b)   (c)

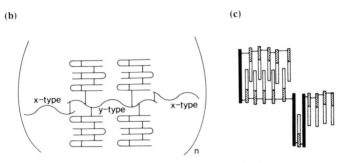

Figure 7.7 Models proposed for the polymeric structure of glutenin. (a) Ewart's linear glutenin model, in which 'concatenations' of polypeptide subunits are linked head-to-tail in random order; (b) The model of Graveland *et al.* in which the high M_r subunits linked head-to-tail form a linear backbone and to which side chain clusters of low M_r subunits are attached; (c) Kasarda's model, in which the low M_r subunits are linked to each other via their C-terminal domains and in which the high M_r subunits are linked into the glutenin polymers via low M_r subunits but not directly to each other. High M_r subunits in this model may be linked to two different low M_r subunits. In all three models the subunits are joined by interchain disulphide bonds. Reproduced with permission from Ewart, ref. 68, Graveland *et al.*, ref. 71 and Kasarda, ref. 72.

force was removed. His model for the subunits is remarkably similar to that proposed recently by Field *et al.*[66] and for which there is now considerable direct experimental evidence.

More recently, Graveland *et al.*[71] proposed a rather different model for polymeric glutenin in which the different types of glutenin polypeptides occur in a repeat unit comprising a *y*-type high M_r subunit attached head-to-tail at each end to two different *x*-type subunits by disulphide bonds (Figure 7.7). Also attached to the *y*-type subunit are four side chain clusters, each containing three low M_r subunits, the M_r of each cluster being approximately 1.2×10^5. In glutenin polymers, these units are linked by disulphide bonds between the N- and C-termini of the *x*-type high M_r subunits resulting in a structure containing a backbone of high M_r subunits with many side chain clusters of low M_r subunits.

Kasarda[72] has proposed yet another model (Figure 7.7) in which the subunits of glutenin are linked by interchain disulphide bonds. It differs from Ewart's model, however, in that low M_r subunits are linked to each other by disulphide bonds in the unrepetitive C-terminal domains (i.e. tail-to-tail) rather than head-to-tail. A further feature of Kasarda's model is that the high M_r subunits are not linked directly to each other but are linked into the gluten polymers via low M_r subunits. The high M_r subunits are envisaged as being linked to different low M_r subunits at their N- and C-termini, resulting in an antiparallel packing of the repetitive domains of the low M_r subunits. Defining the polymeric structure of glutenin, and how it is affected by different glutenin subunit compositions, will be one of the major challenges in the coming years.

The model of Graveland $et\,al.$[71] was based on evidence obtained on the subunit compositions and M_rs of fractions isolated by SDS extraction, ethanol fractionation and gel filtration chromatography. Direct evidence of the location of disulphide bands and of specific associations between subunits was not obtained, however. Other workers have attempted to determine specific associations between subunits by electrophoretic analysis of partially reduced proteins. Lawrence and Payne[73] extracted flour proteins with SDS buffers containing low levels of the reducing agent, 2-mercaptoethanol, and the disulphide compound, cystamine dihydrochloride, and characterised disulphide-linked oligomers in those extracts by one- and two-dimensional electrophoretic techniques. They concluded that oligomer formation amongst the high M_r glutenin subunits showed specificity since only certain subunit combinations were detected. High M_r glutenin subunits encoded on chromosome $1D$ were extensively involved in oligomer formation, and different allelic variants of high M_r subunits encoded on chromosome $1B$ differed in their ability to associate with subunits encoded on other chromosomes. Interestingly, despite the fact that low M_r glutenin subunits comprise around 80% of glutenin, no oligomers were detected that contained low M_r subunits. Dimers comprising in the main an x- and a y-type high M_r glutenin subunit, but also occasionally two x-type subunits, have been detected electrophoretically after partial reduction of SDS-unextractable 'residue' protein of flour.[74] Dimers comprising y-type subunits only were not detected, however.

The occurrence of such dimers and of the high M_r glutenin subunit oligomers detected by Lawrence and Payne[73] is clearly inconsistent with Kasarda's[72] model for glutenin, since, in that model high M_r subunits are linked only to low M_r subunits. As Shewry $et\,al.$[7] have cautioned, however, low concentrations of thiol compounds as used in both these studies can catalyse sulphydryl–disulphide interchange and rearrangement reactions. The possibility that the dimers and higher oligomers that were detected were artefacts of the extraction procedures used cannot be ruled out at present.

Initial attempts to determine directly the location of interpolypeptide chain disulphide bonds, and the specific polypeptides involved, have been reported recently.[75] Disulphide-bonded peptides were isolated chromatographically from a tryptic digest of glutenin. Although three of the peptides isolated were from purothionins, a fourth with the structure

$$\begin{array}{c} \text{CCQQL} \\ |\ | \\ \text{CCQQL} \end{array}$$

was identified, and it was concluded that it was derived through linkage in parallel, i.e. head-to-head, of two y-type high M_r subunits of glutenin, the CCQQL sequence, which occurs at residues 44–48 in such polypeptides. This same sequence occurs also in α- and γ-type gliadins, low M_r subunits of glutenin and wheat endosperm α-amylase inhibitors, however.[7] It is, therefore, not diagnostic of high M_r subunits of glutenin, although in polypeptides other than high M_r subunits of glutenin, the sequence is not flanked by basic amino acid residues and, therefore, the peptide bonds on either side of the sequence in those polypeptides would not be susceptible to tryptic cleavage (H. D. Belitz, personal communication). The relevance to glutenin structure of the disulphide-linked peptide containing it remains to be defined.

7.7 Gluten proteins in relation to breadmaking performance

A major objective of research on wheat proteins has been to define the molecular basis of variation in breadmaking quality and to identify the polypeptide species of greatest importance. A sizeable and rapidly increasing literature now exists on this topic, but as yet the information is not unequivocal.

Of considerable interest has been the possibility of a relationship between the molecular size of glutenin polymers and breadmaking performance, which is a prediction of Ewart's linear glutenin hypothesis.[68–70] Support for such a possibility has come from observations of positive correlations between breadmaking performance and the proportion of glutenin that remains unextractable, on account of its molecular size, in extractants such as 3M urea,[76] dilute acetic acid,[77] SDS/lactic acid solution[78] or SDS solution alone.[79] Similarly, gel filtration chromatography has been used to demonstrate a relationship between the proportion of disulphide-linked glutenin polymers and breadmaking quality using solvents that can extract essentially all the flour or gluten proteins.[80] Moreover, breadmaking performance has also been related to the glutenin polymer size distribution, cultivars of higher breadmaking performance having higher proportions of glutenin polymers of greater molecular size.[81,82] The intrinsic viscosities of glutenin solutions, which can be considered as reflecting molecular size, have also been found to be correlated positively with breadmaking performance.[83]

Much of the interest during the last 10–15 years in defining the molecular basis of breadmaking performance has centred around the effects of specific polypeptides of the gluten protein complex, especially the high M_r subunits of glutenin.[3,8-10] This interest stems from the elegant work of Payne and his colleagues, who have noted correlations between the presence of particular high M_r subunits of glutenin and breadmaking quality in wheats from a number of collections from different countries,[9,84-86] in the progeny from crosses between different cultivars[84,87] and in various types of genetic lines.[88,89]

Numerous studies of this type have now been carried out. In general, the results confirm the importance of high M_r subunits of glutenin in controlling the breadmaking performance of wheat,[3] the different allelic forms of these subunits, which are encoded at three loci on chromosomes 1A, 1B and 1D (termed the *Glu-A1*, *Glu-B1* and *Glu-D1* loci)[8] have been ranked in order of their effects on quality (Table 7.1), and quality scores have been assigned to cultivars on the basis of the allelic subunit types present and their position in these rankings.[10] National collections of wheat vary considerably in their high M_r glutenin subunit compositions, with some collections, such as those from the UK and Germany, on the whole having subunit complements with relatively poor effects on breadmaking performance, whereas others, such as those from Argentina, the USA and Australia, have subunit comp-lements with relatively good effects on quality.[3,10]

The mechanism by which certain high M_r subunits confer better bread-making performance than others remains to be defined. Although some question its importance,[89] the elasticity of gluten is generally considered to be important in relation to its functional role in breadmaking. The discovery that the repetitive domains of high M_r subunits have a β-spiral confor-mational structure that may be intrinsically elastic[66] may therefore be of considerable significance. Primary structure differences between high M_r subunits that affect the β-spiral conformation, and hence glutenin elasticity, could therefore affect breadmaking performance. Comparison of the allelic *y*-type subunits 10 (associated with superior quality) and 12 (associated with inferior quality) indicates that the functional difference may be due to between one and six amino acid substitutions in a short region towards the C-terminal end of the repetitive domain.[90] A more regular β-spiral structure may occur in subunit 10 than in subunit 12, thus conferring superior viscoelastic properties, and hence superior breadmaking properties, to the gluten proteins containing subunit 10.

The disulphide-bonded polymeric structure of glutenin is also important, as discussed above. Differences in the ability of different high M_r subunits to form polymers could be another factor related to the effects of different high M_r subunits on breadmaking performance. As also noted above, however, little is known of the arrangement of subunits within glutein polymers and the effects this might have on glutenin functional properties although it has been shown[91] that high M_r subunit 2, associated with poor quality, differs from its allelic counterpart, subunit 5, associated with good quality, in having a serine residue at position 97 rather than a cysteine residue as in

Table 7.1 Quality scores assigned to individual high M_r subunits of glutenin

Score	Locus		
	Glu-A1	Glu-B1	Glu-D1
4	–	–	5+10
3	1	17+18	–
3	2*	7+ 8	–
3	–	13+16	–
2	–	7+ 9	2+12
2	–	–	3+12
1	null	7	4+12
1	–	6+ 8	–
1	–	20	–

Glutenin subunit alleles consist of null forms (i.e. no subunit produced), single subunit alleles or alleles encoding a pair of subunits*, which are very tightly linked. The quality scores have been assigned mainly on the basis of their effects on values obtained in the SDS-sedimentation test, an indirect measure of breadmaking quality. Scores are ranked from 1 (poor breadmaking quality) to 4 (good breadmaking quality). Reproduced with permission from Payne, ref. 10.

subunit 5. This could influence the potential of the two subunits for polymer formation through interchain disulphide bonds,[7,91] with consequent effects on quality.

Although there has been a tendency to overlook it in favour of qualitative comparisons, at least a part of the variation in glutenin functionality due to different high M_r subunits could be due to quantitative effects on the proportion of high M_r subunit protein present in glutenin. In some of their earliest studies, Payne et al.[84] noted a relationship between the proportion of a certain high M_r subunit (termed subunit 1) present in F2 progeny from a cross between good and poor breadmaking cultivars. Only recently, however, has very much further attention been paid to the question of the effect on functionality of the proportion of glutenin accounted for by high M_r subunits. Both the absolute amounts and relative proportions of high M_r subunits appear to be of importance in this regard.[7,54,55,92–94]

It should be borne in mind that variation among the high M_r subunits of glutenin accounts for only a proportion, albeit, in a number of cases, a substantial proportion, of the variation in breadmaking performance among wheat cultivars. The possible contribution of different gliadin polypeptides to functionality has also been highlighted by Wrigley and co-workers,[3,95–98] although earlier fractionation and reconstitution work had pointed to a role for gliadin in determining the loaf volume potential of flour.[99] Computer-based pattern analyses of (mainly) Australian wheats showed[95–98] that several specific gliadin components were associated with the wheat quality

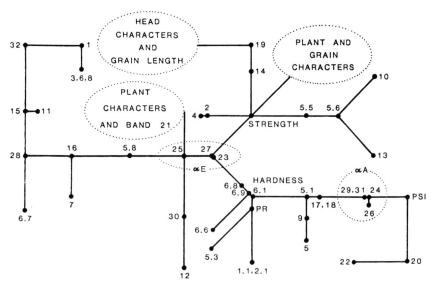

Figure 7.8 Computer-based pattern analysis in the form of a 'minimum spanning tree' showing statistical associations between particular gliadin components (integers from 1–32) and grain quality characteristics for 60 Australian wheat cultivars. The closeness of associations is indicated as distance on the tree. Reproduced with permission from Wrigley, ref. 95.

characteristics (Figure 7.8), grain hardness (see below) and dough strength (an indicator of breadmaking performance). Associations of gliadin bands with quality characteristics have also been noted for Italian[100,101] and French[102] bread wheats. Although the association of grain hardness with specific gliadins mentioned above may have been explicable on the basis of pedigree relationships, the association of specific gliadins and dough strength was not.[97] The conclusion that gliadin polypeptides are associated with both direct and indirect measures of breadmaking performance has been further strengthened through analysis of wheats of diverse quality from a wide range of genetical backgrounds[103] and of advanced breeding lines.[104]

An interesting and potentially important recent observation[105] is that certain α-type gliadin genes contain 'heat-shock elements', i.e. nucleotide sequences associated with the switching on of heat shock genes. This finding may explain, at least in part, the adverse effects on dough properties of high temperatures during grain maturation that have been reported occasionally. Heat-shocked wheats of several cultivars produced grain having higher proportions of gliadin and higher gliadin to glutenin ratios.[105]

The mechanism by which specific gliadin polypeptides might directly influence functional quality has not been established. A similar relationship exists between specific gliadin polypeptides and the pasta-making quality of durum wheat.[105–108] In that case, however, it seems most likely[109,110] that the true explanation of variation in functionality is not the variation in gliadin polypeptides themselves but rather variation in low M_r subunits of glutenin whose structural genes are tightly linked to those for the specific gliadin polypeptides, both being located at the *Gli-B1* locus on the short arm of chromosome 1B. Low M_r subunits of glutenin, γ-gliadins and ω-gliadins are also encoded by genes at equivalent loci on homologous chromosomes 1A and 1D in bread wheats. In a similar manner to the pasta-making quality of durum wheat, associations of specific gliadin components with breadmaking quality could be coincidental, the functionally important associations being with tightly linked low M_r subunits of glutenin and breadmaking performance.[3,9,101]

In fact, the functional role of low M_r subunits of glutenin has received relatively little attention until recently. This is because they have mobilities in the conventional one-dimensional systems of SDS-PAGE similar to those of gliadin polypeptides. Although gliadin and low M_r glutenin polypeptides can be resolved by two-dimensional electrophoretic techniques,[50] the use of such techniques is laborious and unsuited for purposes of screening. Several one-dimensional SDS-PAGE techniques have been developed and refined recently[111–117] for screening the low M_r subunit compositions of cultivar collections, breeding lines, genetical lines etc. (see Figure 7.4, for example). From work already carried out, it appears that variation in low M_r subunits of glutenin may be related to variation in dough properties, and may be as important as variation in high M_r subunits in some cases.[3,112,118] With the advent of these newer techniques for analysing low M_r glutenin subunit compositions, there will undoubtedly be a rapid growth in the literature in the next few years regarding the relationships of the different low M_r glutenin subunit alleles to wheat flour functionality.

7.8 Technological importance of wheat grain endosperm texture

Wheat grain end-use properties are affected markedly by endosperm texture,[11,119] i.e. the hardness or softness of the grain, making this an important raw material quality criterion. Its effects are seen at all levels of the milling process. Energy consumption during grinding, the degree of endosperm/bran separation and hence extraction rate, and the shape and size distribution of flour particles and hence the sieving and transport properties of flour, are all affected. Effects of endosperm texture on flour properties are also carried through to secondary processing, and both processing performance during baking processes and final product quality are affected significantly.

Such is the importance of endosperm texture that it often dictates the type of wheat that will be used for particular products.[11,119] Flours from hard hexaploid (*Triticum aestivum*) wheats are preferred for breadmaking, therefore, while biscuit (cookie) and cake manufacture generally require flours from soft wheats. Semolina from very hard durum wheats, on the other hand, is preferred for pasta making.

In relation to breadmaking, the level of mechanical damage to starch granules during milling is one of the most important effects of endosperm texture on flour properties. Effects of starch damage level on substrate availability for yeast fermentation are perhaps those that are most frequently mentioned in the literature, and can be important.[120] Commercially, however, the effects of starch damage on the water absorbing ability of starch granules is probably of greater significance[121,122] since this can affect dough handling properties, energy consumption during baking, product yield and final product quality characteristics.

Despite its importance, relatively little research has been directed towards defining the biochemical basis for regulation of endosperm texture, although there is an extensive literature[119] describing methods for its assessment, the effects of various edaphic and other factors upon it, and its influence on processing and product characteristics. Definition of the molecular mechanisms that control endosperm texture could lead to new and improved methods for controlling raw material variability at mill intake level, but as with the work described above on the functionality of gluten proteins in relation to breadmaking, it could provide wheat breeders with a precisely defined breeding target and sensitive, small-scale methods by which the property might be selected for at very early stages in the development of new lines with the required quality characteristics.

7.9 Possible mechanisms of endosperm texture variation

Observations on the microstructure of samples of certain UK grown wheat cultivars with widely differing protein contents led Stenvert and Kingswood[123] to propose that the continuity of the (mainly gluten) protein matrix in the endosperm has an important effect on endosperm texture. In those experiments, endosperm texture increased with increasing grain protein content. The relationship between grain protein content and endosperm texture is inconsistent, however, some wheats becoming harder with increasing protein level, some softer and some being unaffected.[119,124] Whether these different effects are due to cultivar or growing conditions is not clear, but nevertheless there appears to be no clear cut evidence that protein level as such controls endosperm texture, although the way the endosperm compacts during the maturation/desiccation phases of grain development, and hence the cohesion of the matrix, could well be significant.[125]

Comparison of the microscopical appearances of flours produced from hard and soft wheats[126,127] has shown that, in the former, the fracture planes tend to be within endosperm cells but aligned with the planes of the endosperm cell walls; they tend not to traverse cells, and thus large 'blocky' flour particles are produced on milling. For soft wheats, on the other hand, the fracture planes tend to traverse endosperm cells and in particular to follow the interface between the starch granule surface and the surrounding matrix of gluten and other proteins and other endosperm components. This results in flours from soft wheats that have a wide range of particle sizes from individual starch granules upwards and also irregularly shaped particles. The cohesion of the endosperm also allows the shearing forces that the grain is subjected to during milling to be transmitted to the starch granules resulting in mechanical damage to the granules, whereas in soft wheats the fact that the starch granule surface/matrix interface provides a plane of weakness minimises mechanical damage to the granules.

Starch granules prepared by non-aqueous techniques from hard wheats have been observed to have much more adherent storage protein on their surfaces than those from soft wheats.[128] Phosphate buffer extractable proteins are also known to be particularly enriched at the peripheries of starch granules.[128,129] Furthermore, micropenetrometer tests on starches and glutens showed no difference in hardness between those components isolated from hard and soft wheats.[128] These observations, as well as those on the way the wheat endosperm fractures, led Simmonds et al.[128,130] to hypothesise that differences in endosperm texture are due to differences in the strength of adhesion between the starch granule surface and the surrounding matrix, but their research failed to identify any specific biochemical component(s) that might be responsible.

7.10 Involvement of a starch granule surface-associated protein in endosperm texture variation

Analysts have known from the early 1990s that well-washed starch preparations contain small amounts of non-carbohydrate materials, notably phosphorus and nitrogen-containing compounds. Phospholipids account for most of the phosphorus and about one third of the nitrogen in the case of wheat starch, but amino acids have also been detected in starch hydroly-sates.[131] The non-lipid nitrogen has been assumed, therefore, to be present as protein, accounting for about 0.15–0.2% of the mass of a well-washed, laboratory prepared prime starch (mostly large lenticular A-type granules), and has been termed 'starch granule protein'.[131]

Until recently, very little work had been done to characterise this protein or investigate its possible technological importance. Work carried out in the last 10 years or so has shown, however, that the starch granule protein is distinct from the bulk of the proteins in flour.[131] Furthermore, two groups of polypeptides can be differentiated, those which are readily extracted with saline or cold SDS solutions, and which are probably associated with the starch granule surface, and those which are extractable only after the granules have been swollen, and which are probably integral components of the granule structure.[131] Some of the latter group are probably enzymes involved in starch biosynthesis, but the physiological roles of the surface proteins have yet to be defined.

Examination by gradient SDS-PAGE of the proteins from starch granule preparations has shown that the amount of a small, M_r 15k starch surface-associated protein varies systematically between hard and soft wheats,[131–133] being present at relatively low levels on hard wheat starches and at relatively high levels on soft wheat starches (Figure 7.9). Starch proteins from several hundred wheat (*T. aestivum*) cultivars from many countries have now been analysed in this way, and the relationship between an intense M_r 15k polypeptide band and endosperm softness and a feint M_r 15k band and endosperm softness is essentially unbroken.[131–133] Other workers have reported similar findings.[134,135] The polypeptide has not been detected on very hard durum wheat starches. These observations provide strong indications, although not proof, of a causal role for this M_r 15k protein, named 'friabilin',[133] in controlling endosperm texture.

A single gene, located on the short arm of chromosome 5D is known to be the major,[136–139] although not the only,[140,141] determinant of endosperm texture variation in wheat. Although called the 'hardness' (*Ha*) gene, softness is, in fact, the dominant trait. This gene is estimated to account for about 65% of that variation in UK wheats,[142] i.e. similar to the proportion of the variation in breadmaking quality in UK wheats accounted for by high M_r subunits of glutenin. Examination of different types of genetical lines of wheat (near isogenic hard and soft pairs, single chromosome substitution lines of a soft wheat containing chromosomes from a hard wheat and homozygous recombinant lines from a cross between a soft wheat and its single chromosome substitution line containing the 5D chromosome from a hard wheat) have also located the 'controlling' gene for friabilin on the short arm of chromosome 5D.[132,133] The controlling gene for friabilin and the *Ha* gene are, in fact, indistinguishable on the basis of analysis of the homozygous recombinant lines. These observations, therefore, reinforce the conclusion that the presence of high levels of friabilin on starch granules and endosperm softness are very closely, and perhaps causally related, although again they do not prove it.

Rye is closely related to wheat, and all cultivars of this cereal known to the author have soft endosperm texture. An analogue of friabilin, although with a slightly greater mobility (apparently lower M_r) on gradient SDS-PAGE, is present in rye.[133,141] Similarly, in many triticale cultivars (a 'man-made' cereal obtained by crossing either tetraploid durum wheat or hexaploid bread wheat with rye) a friabilin analogue is often, although not always present[141] (C. S. Brennan, B. D. Sulaiman and J. D. Schofield, unpublished

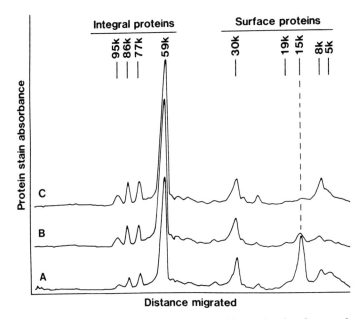

Figure 7.9 Densitometer traces of electrophoregrams of integral and surface starch granule proteins from: (a) the soft milling bread wheat cultivar, Cappelle Desprez; (b) the hard milling bread-wheat cultivar, Bezostaya-1; and (c) the hard-milling single chromosome substitution line, Cappelle Desprez (Bezostaya-5D). Soft milling character is associated with a relatively intense M_r 15k (friabilin) band and hard milling character a relatively weak M_r 15k band. The 5D substitution line was the only line to have a relatively weak M_r 15k band and the only one to have hard milling character suggesting the location on the 5D chromosome of the major controlling genes for both the M_r 15k polypeptide and grain hardness. Polypeptides were separated by SDS-PAGE using a gel with a linear (7.5–20.0%, w/v) acrylamide gradient. Reproduced with permission from Dr. P. Greenwell, Flour Milling and Baking Research Association, Chorleywood, Herts, UK.

results). The triticale analogue, when present, has an electrophoretic mobility similar to that of rye, and presumably originates from the rye parent. Some hexaploid triticale cultivars do not possess the analogue or at least have low levels of it, possibly because of backcrossing to hexaploid bread wheat in the production of 'secondary' triticales with improved agronomic or quality characteristics,[141] resulting in loss of the friabilin-controlling gene.

That the rye analogue of wheat friabilin is also related to endosperm softness is indicated[133] by analysis of starches from single chromosome addition lines of the rye cultivar, King II into the hard wheat cultivar, Holdfast, i.e. Holdfast was the recipient of additions of single chromosome pairs from King II. The 3R chromosome addition line was genetically unstable, therefore unavailable for testing. But addition of all other chromosomes had little effect, either on friabilin analogue band intensity or

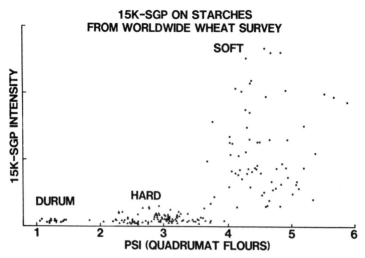

Figure 7.10 Relationship between the milling hardness of wheat as measured by the particle size index (PSI) of Brabender Quadrumat Junior milled flours and friabilin (M_r 15k band) intensity as measured by densitometric scanning of gradient SDS-PAGE separations of starch proteins from a wide range of wheat cultivars from different countries. Friabilin band intensities are expressed in arbitrary densitometer units. Hard milling wheats have low PSI values and soft milling wheats high values. Although starches from soft milling wheats have relatively intense friabilin bands compared with hard wheats, within the soft milling class there is no close relationship between friabilin band intensity and PSI value. Reproduced with permission from Greenwell, ref. 143.

endosperm texture, except for addition of chromosome 5R, which resulted in both an intense friabilin analogue band and soft endosperm texture.

Barley, although exhibiting differences in endosperm texture, which are technologically important, has not been shown conclusively to possess a friabilin analogue.[137]

Although essentially all wheat cultivars classified as soft that have been examined so far have been observed to have a relatively intense friabilin band and all cultivars classified as hard a faint friabilin band, there is considerable variation in endosperm texture within both groups that appears unrelated to friabilin content.[143] This is illustrated by plotting the particle size index (PSI) values, a measure of endosperm texture, against friabilin band intensity as measured by densitometric scanning of stained gradient SDS-PAGE gels for a wide range of cultivars from various countries (Figure 7.10). Somewhat similar, although less extensive, observations have been made by others.[134]

These observations may be explained on the basis of edaphic effects on endosperm texture and/or by the likelihood that genes other than the friabilin/*Ha* gene(s) have significant, although perhaps not the major, effects on endosperm texture, as previously noted (see above). That genes other than that/those associated with the D genome may have effects on endosperm texture is suggested by recent work to examine the starch associated proteins of ancestral forms of wheats and related grasses.[141] Friabilin analogues have been observed in these species even though their genomes are not thought to be closely related to, or to be the precursors of, the D genome of $T.$ $aestivum$.

7.11 Hypothesis for the molecular basis of endosperm texture variation in wheat

These observations are generally consistent with the notion that friabilin plays a causal role in controlling endosperm texture in wheat and some closely related cereals. Friabilin appears to be a starch granule surface component, and is therefore appropriately located to influence adhesion between the granule surface and the surrounding matrix of storage protein and other endosperm components. The available evidence points to its being a product of the Ha gene, and it is present at high levels on soft wheat starches but at low levels on hard wheat starches, which is consistent with softness being the dominant trait.

Simmonds et $al.$[130] considered that differences in endosperm texture might be due to the presence/absence of a component that might act as a 'glue' between the starch granule surface and the surrounding matrix. Greenwell and Schofield[131,132] have modified that hypothesis in the light of the observations described above, however, to suggest that it is friabilin that is the biochemical component responsible for controlling endosperm texture. Rather than acting as a 'glue', friabilin acts as a protein with antiadhesive or 'non-stick' properties, interfering with adhesion between the granule surface and the matrix. On this basis, all wheats would have extremely hard endosperm texture were it not for the presence of low levels of friabilin on hard $T.$ $aestivum$ wheats, and comparatively high levels on soft wheats. The extreme hardness of durum wheats is thus explained by the absence of the D-genome and hence of the Ha/friabilin gene(s), which are carried on chromosome 5D. Little is known at present of the mechanism of adhesion between the starch granule surface and the surrounding matrix or of the mechanism of binding of friabilin to the starch granule surface.

Other recent research[23] has implicated a gene controlling the free (n-hexane extractable) polar lipid levels in flour as also being involved in endosperm texture control. That gene, in fact, has also been found to be indistinguishable from the hardness gene on chromosome 5D on the basis of analysis of the same homozygous recombinant lines referred to above, although, in the case of the lipid work, far fewer lines (20) were analysed than in the friabilin work (100 lines in duplicate). The relationship between

these observations on polar lipid binding and the more extensive observations on friabilin will clearly be important to resolve.

7.12   Application of the knowledge base concerning friabilin

As with definition of the polypeptides that influence breadmaking quality,
one of, if not the, major medium term application of the knowledge base on
friabilin is likely to be in wheat breeding, although applications in raw
material quality control in grain/flour trading may also be possible. Friabilin
may thus provide a molecular marker for this important functional property
of wheat if a suitably rapid and sensitive assay can be developed, e.g. a
quantitative immunoassay.

   Analysis of friabilin levels on starches from F1 hybrids produced by
crossing soft and hard wheats has provided preliminary indications that such
an assay could prove useful.[133] By using four different hard female parents
and one soft male parent and also four different soft female parents and one
hard male parent, four levels of *Ha* gene dosage were available in the F1
hybrids. Friabilin on the starches was measured semi-quantitatively by
densitometric scanning of stained gradient SDS-PAGE gels. The results
showed that friabilin levels were lowest in the hard parent cultivars and
highest in the soft parent cultivars, as expected, with intermediate levels
being present in the two sets of F1 hybrids, with the hybrids from the crosses
involving soft female parents being clearly distinguishable from those
involving hard female parents. Both the friabilin levels and PSI values varied
as expected for the different sets of parent cultivars and the F1 hybrids.
Assuming that friabilin levels in intact endosperms do, in fact, reflect the
levels of friabilin associated with the isolated starches, friabilin does appear
as though it may be a useful molecular marker of endosperm texture
differences.

7.13   Concluding remarks

The foregoing discussion indicates clearly that a great deal of progress has
been made in recent years in identifying those wheat proteins that have
important functional roles in breadmaking particularly, but also more
recently in milling. A great deal has been learned about the structures of the
individual polypeptides that are important to breadmaking, although the
knowledge base in relation to components controlling milling quality is not
so well advanced as yet. The role of the low M_r subunits of glutenin in
controlling breadmaking quality is only just beginning to be unravelled, but
this area will undoubtedly advance rapidly in the next few years now that
techniques are available for screening these components efficiently.

Despite the impressive advances which have come from thorough research in a number of different although related disciplines, much still remains to be done to understand the molecular basis of the functionality of wheat proteins, whether in relation to breadmaking or milling. Elucidating the polymeric structure of glutenin will be particularly challenging, but only when this is done, and the nature of the individual polypeptide components of both the glutenin and gliadin complexes known fully, will we be able to understand in precise terms how these components bring about their functional effects. This knowledge base will also be needed to understand in detail the non-covalent interactions between the gluten proteins themselves and with other components of flour and dough, such as flour lipids, emulsifier ingredients, baking fat, water and the redox systems present endogenously in flour and added in the form of redox improvers.

We have learned a great deal about the nature and functionality of the protein components of wheat, but much of this research is oriented towards producing new improved cultivars of wheat through better defined breeding targets and, no doubt, in the future through genetic engineering to alter the types or amounts of different polypeptides present. No doubt this will benefit end-users of grain, such as millers and bakers in the long term. Relatively little research has so far been directed towards defining the behaviour of these proteins during processes such as breadmaking, however, and to understanding the often fairly small but commercially significant variation that can and does occur in processing performance between ostensibly similar lots of wheat. In fact, this is important for wheat breeders as well as processors so that they are provided with a much clearer definition than exists at present of what functional properties are required in wheat for specific end-uses. If significant progress is to be made in this direction in future, it will require some change in perspective on the part of research scientists working in this area and a much clearer and exact definition from those working in commercial environments of the properties they require in wheats and the nature of the processing problems they face.

References

1. Schofield J. D. and Booth, M. R. (1983) Wheat proteins and their technological significance. In *Developments in Food Proteins–2*, (ed. B. J. F. Hudson) Applied Science, Barking, Essex, UK, pp. 1–65.
2. Wrigley, C. W. and Bietz, J. A. (1988) Proteins and amino acids. In *Wheat Chemistry and Technology*, 3rd edn, vol. 1, (ed. Y. Pomeranz) American Association of Cereal Chemists, St. Paul, Minnesota, pp. 159–275.
3. MacRitchie, F., du Cros, D. L. and Wrigley, C. W. (1990) Flour polypeptides related to wheat quality. *Adv. Cereal Sci. Technol.*, **10**, 79–145.
4. Shewry, P. R. and Miflin, B. J. (1955) Seed storage proteins of economically important cereals. *Adv. Cereal Sci. Technol.*, **7**, 1–84.
5. Shewry, P. R., Halford, N. G. and Tatham, A. S. (1989) The high molecular weight subunits of wheat, barley and rye: Genetics, molecular biology, chemistry and role in

wheat gluten structure and functionality. *Oxford Surveys Plant Molec. Cell Biol.*, **6**, 163–219.

6. Tatham, A. S., Shewry, P. R. and Belton, P. S. (1990) Structural studies of cereal prolamins, including wheat gluten. *Adv. Cereal Sci. Technol.*, **10**, 1–78.

7. Shewry, P. R., Halford, N. G. and Tatham, A. S. (1992) High molecular weight subunits of wheat glutenin. *J. Cereal Sci.*, **15**, 115–119.

8. Payne, P. I., Holt, L. M., Jackson, E. A. and Law, C. N. (1984) Wheat storage proteins: their genetics and their potential for manipulation by plant breeding. *Philos. Trans R. Soc. London, Ser. B*, **304**, 359–371.

9. Payne, P. I. (1987) Genetics of wheat storage proteins and the effect of allelic variation on breadmaking quality. *Ann. Rev. Plant Physiol.*, **38**, 141–153.

10. Payne, P. I. (1987) The genetical basis of breadmaking quality in wheat. *Aspects Appl. Biol.*, **15**, 79–90.

11. Kent, N. L. (1984) *Technology of Cereals*, 3rd edn, Pergamon Press, Oxford.

12. Schofield, J. D. and Greenwell, P. (1987) Wheat starch granule proteins and their technological significance. In *Cereals in a European Context*, (ed. I. D. Morton) VCH Verlagsgesellschaft mbH, Weinheim and Ellis Horwood, Chichester, pp. 407–420.

13. Finney, K. F. (1943) Fractionating and reconstituting techniques as tools in wheat flour research. *Cereal Chem.*, **20**, 381–396.

14. Finney, K. F. and Barmore, M. A. (1948) Loaf volume and protein content of hard winter and spring wheats. *Cereal Chem.*, **25**, 291–312.

15. Bushuk, W. (1985) Wheat flour proteins: Structure and role in breadmaking, in *Analyses as Practical Tools in the Cereal Field*, (ed. K. M. Fjell) Norwegian Grain Corp., Oslo, pp. 187–198.

16. Booth, M. R. and Melvin, M. A. (1979) Factors responsible for the poor breadmaking quality of high yielding European wheat. *J. Sci. Food Agric.*, **30**, 1057–1064.

17. MacRitchie, F. (1978) Differences in baking quality between wheat flours. *J. Food Technol.*, **13**, 187–194.

18. MacRitchie, F. (1985) Studies on the methodology for fractionation and reconstitution of wheat flours. *J. Cereal Sci.*, **3**, 221–230.

19. Hoseney, R. C., Finney, K. F., Shogren, M. D. and Pomeranz, Y. (1969) Functional (breadmaking) and biochemical properties of wheat flour components. II. Role of water solubles. *Cereal Chem.*, **46**, 117–125.

20. Chung, O. K., Pomeranz, Y. and Finney, K. F. (1982) Relation of polar lipid content to mixing requirement and loaf volume potential of hard red winter wheat flour. *Cereal Chem.*, **59**, 14–20.

21. Zawistowska, U., Bekes, F. and Bushuk, W. (1984) Intercultivar variations in lipid content, composition and distribution and their relation to baking quality. *Cereal Chem.*, **61**, 527–531.

22. Bekes, F., Zawistowska, U., Zillman, R. R. and Bushuk, W. (1986) Relationship between lipid content and composition and loaf volume of twenty-six common spring wheats. *Cereal Chem.*, **63**, 327–331.

23. Morrison, W. R., Law, C. N., Wylie, L. J., Coventry, A. M. and Seekings, J. (1989) Effect of group 5 chromosomes on the free polar lipids and breadmaking quality of wheat. *J. Cereal Sci.*, **9**, 41–51.

24. Huebner, F. and Wall, J. S. (1976) Fractionation and quantitative differences of glutenin from wheat varieties varying in baking quality. *Cereal Chem.*, **53**, 258–269.

25. Byers, M., Miflin, B. J. and Smith, S. J. (1983) A quantitative comparison of the extraction of protein fractions from wheat grain by different solvents, and of the polypeptide and amino acid composition of the alcohol-soluble proteins. *J. Sci. Food Agric.*, **34**, 447–462.

26. Webb, T., Heaps, P. W. and Coppock, J. B. M. (1971) Protein quality and quantity: a rheological assessment of their relative importance in breadmaking. *J. Food Technol.*, **6**, 47–62.

27. Pernollet, J. C. and Mossé, J. (1983) Structure and location of legume and cereal seed storage proteins. In *Seed Proteins* (ed. J. Daussant, J. Mossé and J. G. Vaughan) Academic Press, NY and London, pp. 155–191.

28. Wall, J. S. (1979) The role of wheat proteins in determining baking quality. In *Recent Advances in the Biochemistry of Cereals*, (ed. D. L. Laidman and R. W. Wyn-Jones)

Phytochem. Soc. Eur. Symp. Ser. No. 16, Academic Press, London, NY and San Francisco, pp. 275–311.

29. Orth, R. A. and Bushuk, W. (1972) A comparative study of the proteins of wheats of diverse baking quality. *Cereal Chem.*, **49**, 268–275.

30. Orth, R. A. and O'Brien, L. (1976) A new biochemical test of dough strength of wheat flour. *J. Aust. Inst. Agric. Sci.*, **42**, 122–124.

31. Payne, P. I. and Corfield, K. D. (1979) Subunit composition of wheat glutenin proteins isolated by gel filtration in a dissociating medium. *Planta*, **145**, 83–88.

32. Bietz, J. A. and Wall, J. S. (1980) Identity of high molecular weight gliadin and ethanol-soluble glutenin subunits of wheat: Relation to gluten structure. *Cereal Chem.*, **57**, 415–421.

33. Bottomley, R. C., Kearns, H. F. and Schofield, J. D. (1982) Characterisation of wheat flour and gluten proteins using buffers containing sodium dodecyl sulphate. *J. Sci. Food Agric.*, **33**, 481–491.

34. Field, J. M. Shewry, P. R., Burgess, S. R., Forde, J., Parmar, S. and Miflin, B. J. (1983) The presence of high molecular weight aggregates in protein bodies of developing endosperms of wheat and other cereals. *J. Cereal Sci.*, **1**, 33–41.

35. Payne, P. I., Holt, L. M., Burgess, S. R. and Shewry, P. R. (1986) Characterisation by two-dimensional gel electrophoresis of the protein components of protein bodies isolated from the developing endosperm of wheat (*Triticum aestivum*). *J. Cereal Sci.*, **4**, 217–223.

36. Miflin, B. J., Field, J. M. and Shewry, P. R. (1983) Cereal storage proteins and their effect on technological properties. In *Seed Proteins*, (ed. J. Daussant, J. Mossé and J. G. Vaughan) Academic Press, London, New York and San Francisco, pp. 255–319.

37. Shewry, P. R., Tatham, A. S., Forde, J., Kreis, M. and Miflin, B. J. (1986) The classification and nomenclature of wheat gluten protein: A reassessment. *J. Cereal Sci.*, **4**, 97–106.

38. Schofield, J. D. (1986) Flour proteins: Structure and functionality in baked products, in *Chemistry and Physics of Baking*, (eds J. M. V. Blanshard, P. J. Frazier and T. Galliard) Roy. Soc. Chem., London, pp. 14–29.

39. Payne, P. I., Holt, L. M., Jarvis, M. G. and Jackson, E. A. (1985) Two-dimensional fractionation of the endosperm proteins of bread wheat (*Triticum aestivum*): Biochemical and genetic studies. *Cereal Chem.*, **62**, 317–326.

40. Jones, R. W., Taylor, N. W. and Senti, F. R. (1959) Electrophoresis and fractionation of wheat gluten. *Arch. Biochem. Biophys*, **84**, 363–376.

41. Elton, G. A. H. and Ewart, J. A. D. (1960) Starch gel electrophoresis of wheat proteins. *Nature*, **187**, 600–601.

41. Woychik, J. H., Boundy, J. A. and Dimler, R. J. (1961) Starch gel electrophoresis of wheat gluten proteins with concentrated urea. *Arch. Biochem. Biophys.*, **94**, 477–482.

43. Wrigley, C. W., Autran, J.-C. and Bushuk, W. (1982) Identification of cereal varieties by gel electrophoresis of the grain proteins. *Adv. Cereal Sci. Technol.*, **5**, 211–259.

44. Booth, M. R. and Ewart, J. A. D. (1969) Studies on four components of wheat gliadins. *Biochim. Biophys. Acta*, **181**, 226–233.

45. Charbonnier, L. (1974) Isolation and characterization of omega-gliadin fractions. *Biochim. Biophys. Acta*, **359**, 142–151.

46. Kasarda, D. D., Autran, J.-C., Lew, E. J.-L., Nimmo, C. C. and Shewry, P. R. (1983) N-terminal amino acid sequences of ω-gliadins and ω-secalins: Implications for the evolution of prolamin genes. *Biochim. Biophys. Acta*, **747**, 138–150.

47. Wrigley, C. W. and Shepherd, K. W. (1973) Electrofocusing of grain proteins from wheat genotypes. *Ann. NY Acad. Sci.*, **209**, 154–162.

48. Wrigley, C. W., Autran, J.-C. and Bushuk, W. (1982) Identification of cereal varieties by gel electrophoresis of the grain proteins. *Adv. Cereal Sci. Technol.*, **5**, 211–259.

49. Ellis, J. R. S. (1984) The cereal grain trade in the United Kingdom: The problem of cereal variety. *Philos. Trans. R. Soc. London, Ser B.*, **304**, 395–407.

50. Jackson, E. A., Holt, L. M. and Payne, P. I. (1983) Characterisation of high molecular weight gliadin and low molecular weight glutenin subunits of wheat endosperm by two-dimensional electrophoresis and the chromosomal location of their controlling genes. *Theor. Appl. Genet.*, **66**, 29–37.

51. Bunce, N. A. C., White, R. P. and Shewry, P. R. (1985) Variation in estimates of molecular weights of cereal prolamins by SDS-PAGE. *J. Cereal Sci.*, **3**, 131–142.

52. Field, J. M., Shewry, P. R., Miflin, B. J. and March, J. F. (1982) The purification and characterisation of homologous high molecular weight storage proteins from grain of wheat, barley and rye. *Theor. Appl. Genet.*, **62**, 329–336.
53. Anderson, O. D., Halford, N. G., Forde, J., Yip, R., Shewry, P. R. and Greene, F. L. (1988) Structure and analysis of the high molecular weight glutenin genes from *Triticum aestivum* L cv. Cheyenne, in *Proceedings 7th International Wheat Genetics Symposium*, (ed T. E. Miller and R. M. D. Koebner) IPSR, Cambridge, pp. 699–704.
54. Sutton, K. H. (1991) Qualitative and quantitative variation among high molecular weight subunits of glutenin detected by reversed-phase high performance liquid chromatography. *J. Cereal Sci.*, **14**, 25–34.
55. Seilmeier, W., Belitz, H.-D. and Wieser, H. (1991) Separation and quantitative determination of high molecular weight subunits of glutenin from different wheat varieties and genetic variants of the variety Sicco. *Z. Lebensm. Unters Forsch.*, **192**, 124–129.
56. Bietz, J. A. and Wall, J. S. (1980) Identity of high molecular weight gliadin and ethanol-soluble glutenin subunits of wheat: Relation to gluten structure. *Cereal Chem.*, **57**, 415–421.
57. Kasarda, D. D., Bernardin, J. E. and Nimmo, C. C. (1976) Wheat proteins. *Adv. Cereal Sci. Technol.*, **1**, 158–236.
58. Payne, P. I., Holt, L. M. and Law, C. N. (1981) Structural and genetic studies on the high molecular weight subunits of wheat glutenin. Part 1. Allelic variation in subunits amongst varieties of wheat (*Triticum aestivum*). *Theor. Appl. Genet.*, **60**, 229–236.
59. Kasarda, D. D., Bernardin, J. E. and Gaffield, W. (1968) Circular dichroism and optical rotary dispersion of α-gliadin. *Biochemistry*, **7**, 3950–3957.
60. Tatham, A. S. and Shewry, P. R. (1985) The conformation of wheat gluten proteins. The secondary structures and thermal stabilities of α-, β-, γ- and ω-gliadins. *J. Cereal. Sci.*, **3**, 103–113.
61. Tatham, A. S., Miflin, B. J. and Shewry, P. R. (1985) The β-turn conformation in wheat gluten proteins. Relationships to gluten elasticity. *Cereal Chem.*, **62**, 405–412.
62. Tatham, A. S., Field, J. M., Smith, S. J. and Shewry, P. R. (1987). The conformation of wheat gluten proteins. 2. Aggregated gliadins and LMW subunits of glutenin. *J. Cereal Sci.*, **5**, 203–214.
63. Purcell, J. M., Kasarda, D. D. and Wu, C.-S. C. (1988) Secondary structures of wheat α- and ω-proteins: Fourier transform infra-red spectroscopy. *J. Cereal Sci.*, **7**, 21–32.
64. Tatham, A. S., Mason, P. and Popineau, Y. (1990) Conformational studies of peptides derived by the enzymic hydrolysis of a gamma-type gliadin. *J. Cereal Sci.*, **11**, 11–13.
65. Tatham, A. S., Shewry, P. R. and Miflin, B. J. (1984) Wheat gluten elasticity: A similar molecular basis to elastin. *FEBS Lett.*, **177**, 205–208.
66. Field, J. M., Tatham, A. S. and Shewry, P. R. (1987) The structure of a high M_r subunit of durum wheat (*T. durum*) gluten. *Biochem. J.*, **247**, 215–221.
67. Miles, M. J., Carr, H. J., McMaster, T., Belton, P. S., Morris, V. J., Field, J. M., Shewry, P. R. and Tatham, A. S. (1991) Scanning, tunnelling microscopy of a wheat seed storage protein reveals details of an unusual supersecondary structure. *Proc. Natl. Acad. Sci. USA*, **88**, 68–71.
68. Ewart, J. A. D. (1968) A hypothesis for the structure and rheology of glutenin, *J. Sci. Food Agric.*, **19**, 617–623.
69. Ewart, J. A. D. (1977) Re-examination of the linear glutenin hypothesis. *J. Sci. Food Agric.*, **28**, 191–199.
70. Ewart, J. A. D. (1979) Glutenin structure. *J. Sci. Food Agric.*, **30**, 482–492.
71. Graveland, A., Bosveld, P., Lichtendonk, W. J., Marseille, J. P., Moonen, J. H. E. and Scheepstra, A. (1985) A model for the molecular structure of the glutenins from wheat flour. *J. Cereal Sci.*, **3**, 1–16.
72. Kasarda, D. D. (1989) Glutenin structure in relation to wheat quality, in *Wheat is Unique*, (ed. Y. Pomeranz) American Association of Cereal Chemists, St. Paul, MN, USA, pp. 277–302.
73. Lawrence, G. J. and Payne, P. I. (1983) Detection by gel electrophoresis of oligomers formed by the association of high molecular weight glutenin protein subunits of wheat endosperm. *J. Exp. Bot.*, **34**, 254–267.
74. Werner, W. E., Adelsteins, A. E. and Kasarda, D. D. (1992) Composition of high

molecular weight glutenin subunit dimers formed by partial reduction of residue protein. *Cereal Chem.*, **69**, 535–541.

75. Köhler, P., Belitz, H.-D. and Weiser, H. (1991) Disulphide bonds in wheat gluten: isolation of a cystine peptide from glutenin. *Z. Lebensm. Unters. Forsch.*, **192**, 234–239.

76. Pomeranz, Y. (1965) Dispersibility of wheat proteins in aqueous urea solutions – new parameter to evaluate breadmaking potentialities of wheat flour. *J. Sci. Food Agric.*, **16**, 586–593.

77. Orth, R. A. and Bushuk, W. (1972) A comparative study of the proteins of wheats of diverse baking qualities. *Cereal Chem.*, **49**, 268–275.

78. Axford, D. W. E., McDermott, E. E. and Redman, D. G. (1979) Note on the sodium dodecyl sulphate test and breadmaking quality: comparison with Pelshenke and Zeleny tests. *Cereal Chem.*, **56**, 582–584.

79. Moonen, J. H. E., Scheepstra, A. and Graveland, A. (1983) The positive effects of the high molecular weight subunits 3 + 10 and 2* of glutenin on the breadmaking quality of wheat cultivars. *Euphytica*, **32**, 735–742.

80. Field, J. M., Shewry, P. R. and Miflin, B. J. (1983) Solubilization and characterization of wheat gluten proteins: Correlations between the amounts of aggregated proteins and baking quality. *J. Sci. Food Agric.*, **34**, 370–377.

81. Huebner, F. and Wall, J. S. (1976) Fractionation and quantitative differences of glutenin from wheat varieties varying in baking quality. *Cereal Chem.*, **53**, 258–269.

82. Bottomley, R. C., Kearns, H. F. and Schofield, J. D. (1982) Characterisation of wheat flour and gluten proteins using buffers containing sodium dodecyl sulphate. *J. Sci. Food Agric.*, **33**, 481–491.

83. Ewart, J. A. D. (1980) Loaf volume and the intrinsic viscosity of glutenin. *J. Sci. Food Agric.*, **31**, 1323–1336.

84. Payne, P. I., Corfield, K. G. and Blackman, J. A. (1979) Identification of a high molecular weight subunit of glutenin whose presence correlates with breadmaking quality in wheats of related pedigree. *J. Sci. Food Agric.*, **55**, 153–159.

85. Payne, P. I., Nightingale, M. A., Krattiger, A. F. and Holt, L. M. (1987) The relationship between HMW glutenin subunit composition and the breadmaking quality of British-grown wheat varieties. *J. Sci. Food Agric.*, **40**, 51–65.

86. Payne, P. I., Holt, L. M., Krattiger, A. F. and Carrillo, J. M. (1988) Relationships between seed quality characteristics and HMW glutenin subunit composition determined using wheats grown in Spain. *J. Cereal Sci.*, **7**, 229–235.

87. Payne, P. I., Corfield, K. G., Holt, L. M. and Blackman, J. A. (1981) Correlations between the inheritance of certain high molecular weight subunits of glutenin and breadmaking quality in progenies of six crosses of bread wheat. *J. Sci. Food Agric.*, **32**, 51–60.

88. Rogers, W. J., Payne, P. I., Seekings, J. A. and Sayers, E. J. (1991) Effect on breadmaking quality of x-type and y-type high molecular weight subunits of glutenin. *J. Cereal Sci.*, **14**, 209–222.

89. Bloksma, A. H. (1990) Rheology of the breadmaking process. *Cereal Foods World*, **35**, 228–235.

90. Goldsborough, A. P., Bulleid, N. J., Freedman, R. B. and Flavell, R. B. (1989) Conformational differences between two wheat (*Triticum aestivum*) 'high molecular weight' glutenin subunits are due to a short region containing six amino acid differences. *Biochem. J.*, **263**, 837–842.

91. Green, F. C., Anderson, R. E., Yip, R. E., Halford, N. G., Malpica Romero, J.-M. and Shewry, P. R. (1988) Analysis of possible quality-related sequence variations in the 1D glutenin high molecular weight subunit genes of wheat, in *Proceeding 7th International Wheat Genetics Symposium*, Vol 1, (ed. T. E. Miller and R. M. D. Koebner) IPSR, Cambridge, UK, pp. 735–740.

92. Sutton, K. H., Hay, R. L. and Griffin, W. B. (1989) Assessment of the potential bread baking quality of New Zealand wheats by RP-HPLC of glutenins. *J. Cereal Sci.*, **10**, 113–122.

93. Marchylo, B. A., Luckow, O. M. and Kruger, J. E. (1992) Quantitative variation in high molecular weight glutenin subunit 7 in some Canadian wheats. *J. Cereal Sci.*, **15**, 29–38.

94. Kolster, P., Krechting, C. F. and van Gelder, W. M. J. (1992) Quantification of individual

high molecular weight subunits of wheat glutenin using SDS-PAGE and scanning densitometry. *J. Cereal Sci.*, **15**, 49–62.

95. Wrigley, C. W. (1980) The genetic and chemical significance of varietal differences in gluten composition. *Ann. Technol. Agric.*, **29**, 213–227.

96. Wrigley, C. W., Robinson, P. J. and Williams, W. T. (1981) Association between electrophoretic patterns of gliadin proteins and quality characteristics of wheat cultivars. *J. Sci. Food Agric.*, **32**, 433–442.

97. Wrigley, C. W., Robinson, P. J. and Williams, W. T. (1982) Relationships between Australian wheats on the basis of pedigree, grain protein composition and grain quality in wheat. *Aust. J. Agric. Res.*, **33**, 419–427.

98. Wrigley, C. W., Lawrence, G. J. and Shepherd, K. W. (1982) Association of glutenin subunits with gliadin composition and grain quality in wheat. *Aust. J. Plant Physiol.*, **9**, 15–30.

99. Hoseney, R. C., Finney, K. F., Pomeranz, Y. and Shogren, M. D. (1969) Functional (breadmaking) and biochemical properties of wheat flour components. IV. Gluten protein fractionation by solubilizing in 70% ethyl alcohol and in dilute lactic acid. *Cereal Chem.*, **46**, 495–518.

100. Pogna, N. E., Boggini, G., Corbellini, M., Cattaneo, M. and Dal Belin Peruffo, A. (1982) Association between gliadin electrophoretic bands and quality in common wheat. *Can. J.Plant Sci.*, **62**, 913–918.

101. Dal Belin Peruffo, A., Pogna, N. E., Tealdo, E., Tutta, C. and Alubzio, A. (1985) Isolation and partial characterisation of gamma-gliadins 40 and 43.5 associated with quality in common wheat. *J. Cereal Sci.*, **3**, 355–362.

102. Campbell, W. P., Wrigley, C. W., Cressey, P. J. and Slack, C. R. (1989) Statistical correlations between quality attributes and grain protein composition for 71 hexaploid wheats used as breeding parents. *Cereal Chem.*, **64**, 293–299.

103. Cressey, P. J., Campbell, W. P., Wrigley, C. W. and Griffin, W. B. (1987) Statistical correlations between quality attributes and grain protein composition for 60 advanced lines of crossbred wheat. *Cereal Chem.*, **64**, 299–301.

104. Blumenthal, C. S., Batey, I. L., Bekes, F., Wrigley, C. W. and Barlow, E. W. R. (1990) Gliadin genes contain heat-shock elements: Possible relation to heat-induced changes in grain quality. *J. Cereal Sci.*, **11**, 185–188.

105. Damidaux, R., Autran, J.-C., Grignac, P. and Feillet, P. (1978) Mise en evidence de relations applicable in selection entre l'electrophoregramme des gliadines et les proprietes viscoelastiques du gluten de *Triticum durum* Desf. *C. R. Seances Acad. Sci. Ser. D*, **287**, 701–704.

106. Damidaux, R., Autran, J.-C. and Feillet, P. (1980) Gliadin electrophoregrams and measurements of gluten viscoelasticity in durum wheats. *Cereal Foods World*, **25**, 754–756.

107. Kosmolak, F. G., Dexter, J. E., Matsuo, R. R., Leisle, D. and Marchylo, B. A. (1980) A relationship between durum wheat quality and gliadin electrophoregrams. *Can. J. Plant Sci.*, **60**, 427–432.

108. du Cros, D. L., Wrigley, C. W. and Hare, R. A. (1982) Prediction of durum wheat quality from gliadin protein composition. *Aust. J. Agric. Res.*, **33**, 429–442.

109. Payne, P. I., Jackson, E. A. and Holt, L. M. (1984) The association between gamma-gliadin 45 and gluten strength in durum wheat varieties: a direct causal effect or the result of genetic linkage? *J. Cereal Sci.*, **2**, 73–81.

110. Pogna, N. E., Lafiandra, D., Feillet, P. and Autran, J.-C. (1988) Evidence for a direct causal effect of low molecular weight subunits of glutenins on gluten viscoelasticity in durum wheats. *J. Cereal Sci.*, **7**, 211–214.

111. Singh, N. K. and Shepherd, K. W. (1988) Linkage mapping of genes controlling endosperm storage proteins in wheat. 1. Genes on the short arms of group 1 chromosomes. *Theor. Appl. Genetics*, **75**, 628–641.

112. Gupta, R. B. and Shepherd, K. W. (1989) Low molecular weight glutenin subunits in wheat: their variation, inheritance, and association with breadmaking quality, in *Proceedings 7th International Wheat Genetics Symposium*, (ed. T. E. Miller and R. M. D. Koebner), I.P.S.R., Cambridge, pp. 943–949.

113. Graybosch, R. A. and Morris, R. (1990) An improved SDS-PAGE method for the analysis of wheat endosperm storage proteins. *J. Cereal Sci.*, **11**, 201–212.

114. Khelifi, D. and Branlard, G. (1991) A new two-step electrophoresis method for analysing gliadin polypeptides and high- and low molecular weight subunits of glutenin of wheat. *J. Cereal Sci.*, **13**, 41–48.
115. Gupta, R. B. and MacRitchie, F. (1991) A rapid one-step one-dimensional SDS-PAGE procedure for analysis of subunit composition of glutenin in wheat. *J. Cereal Sci.*, **14**, 105–110.
116. Singh, N. K., Shepherd, K. W. and Cornish, G. B. (1991) A simplified SDS-PAGE procedure for separating LMW subunits of glutenin. *J. Cereal Sci.*, **14**, 203–208.
117. Zhen, Z. and Mares, D. (1992) A simple extraction and one-step SDS-PAGE system for separating HMW and LMW glutenin subunits of wheat and high molecular weight proteins of rye. *J. Cereal Sci.*, **15**, 63–78.
118. Gazanhes, V., Morel, M. H. and Autran, J.-C. (1991) The low molecular weight glutenin composition of French bread wheats and its effect on dough properties. *Cereal Foods World*, **36**, 723, Abs. 277.
119. Pomeranz, Y. and Williams, P. C. (1990) Wheat hardness: Its genetic, structural and biochemical background, measurement and significance. *Adv. Cereal Sci. Technol.*, **10**, 471–548.
120. Farrand, E. A. (1969) Starch damage and *alpha*-amylase as basis for mathematical models relation to flour water absorption. *Cereal Chem.*, **46**, 103–116.
121. Stevens, D. J. (1987) Water absorption of flour, in *Cereals in a European Context* (ed I. D. Morton), VCH Verlagsgesellschaft mbH, Weinheim and Ellis Horwood, Chichester, pp. 273–284.
122. Evers, A. D. and Stevens, D. J. (1985) Starch damage. *Adv. Cereal Sci. Technol.*, **7**, 321–349.
123. Stenvert, N. L. and Kingswood, K. (1977) The influence of the physical structure of the protein matrix on wheat hardness. *J. Sci. Food Agric.*, **28**, 11–19.
124. Pomeranz, Y., Peterson, C. J. and Mattern, P. J. (1985) Hardness of winter wheats grown under widely different climatic conditions. *Cereal Chem.*, **62**, 463–467.
125. MacRitchie, F. (1980) Physiochemical aspects of some problems in wheat research. *Adv. Cereal Sci. Technol.*, **3**, 271–326.
126. Greer, E. N. and Hinton, J. J. C. (1950) The two types of wheat endosperm. *Nature*, **165**, 746–748.
127. Hoseney, R. C. and Seib, P. A. (1973) Structural differences in hard and soft wheats. *Baker's Digest*, **47**, 26 and 56.
128. Barlow, K. K., Buttrose, S. M., Simmonds, D. H. and Vesk, K. (1973) The nature of the starch protein interface in wheat endosperm. *Cereal Chem.*, **50**, 443–454.
129. Barlow, K. K., Simmonds, D. H. and Kenrick, K. G. (1973) The localization of water soluble proteins in the wheat endosperm as revealed by fluorescent antibody techniques. *Experientia*, **29**, 229–235.
130. Simmonds, D. H., Barlow, K. K. and Wrigley, C. W. (1973) The biochemical basis of grain hardness in wheat. *Cereal Chem.*, **50**, 553–562.
131. Schofield, J. D. and Greenwell, P. (1987) Wheat starch granule proteins and their technological significance. In *Cereals in a European Context*, (ed. I. D. Morton) VCH Verlagsgesellschaft mbH, Weinheim and Ellis Horwod, Chichester, pp. 407–420.
132. Greenwell, P. and Schofield, J D. (1986) A starch granule protein associated with endosperm softness in wheat. *Cereal Chem.*, **63**, 379–380.
133. Greenwell, P. and Schofield, J. D. (1989) The chemical basis of grain hardness and softness, in *Wheat End-Use Properties. Proc ICC '89 Symposium*, (ed H. Salovaara) University of Helsinki, pp. 59–72.
134. Glenn, G. M. and Saunders, R. M. (1990) Physical and structural properties of wheat endosperm associated with grain texture. *Cereal Chem.*, **67**, 176–182.
135. Bakhella, M., Hoseney, R. C. and Lockhart, G. L. (1990) Hardness of Moroccan wheats. *Cereal Chem.*, **67**, 246–250.
136. Symes, K. J. (1965) The inheritance of grain hardness in wheat as measured by the particle size index. *Aust. J. Agric. Res.*, **16**, 113–123.
137. Symes, K. J. (1969) The influence of a gene causing hardness on the milling and baking qualities of two wheats. *Aust. J. Agric. Res.*, **20**, 971–979.
138. Law, C. N., Young, C. F., Brown, J. W. S., Snape, J. W. and Worland, A. J. (1978) The study of grain protein control in wheat using whole chromosome substitution lines,

in *Seed Improvement by Nuclear Techniques*, Intl. Atomic Energy Agency, Vienna, pp. 483–502.

139. Law, C. N. and Krattiger, A. F. (1987) Genetics of grain quality in wheat. In *Cereals in a European Context*, (ed. I. D. Morton) VCH Verlagsgesellschaft, Weinheim and Ellis Horwood, Chichester, pp. 33–47.

140. Konzak, C. F. (1977) Genetic control of the content, amino acid composition, and processing properties of proteins in wheat. *Adv. Genetics*, **19**, 407–582.

141. Morrison, W. R., Greenwell, P., Law, C. N. and Sulaiman, B. D. (1992) Occurrence of friabilin, a low molecular weight protein associated with grain softness, on starch granules isolated from some wheats and related species. *J. Cereal Sci.*, **15**, 143–149.

142. Krattiger, A. F. (1989) The genetics and biochemistry of breadmaking quality in wheat (*Triticum aestivum* L.) *PhD thesis*, University of Cambridge.

143. Greenwell, P. (1987) Wheat starch granule proteins and their technological significance, in *Proceedings 37th Australia Cereal Chem. Conference*, (ed. L. Murray) Roy. Aust. Chem. Inst., Melbourne, pp. 100–103.

8 Wheat carbohydrates: structure and functionality*

B. L. D'APPOLONIA and P. RAYAS-DUARTE

8.1 Introduction

Wheat carbohydrates have been studied extensively over the years in terms of structure and functionality as related to a particular end-product. Perhaps the majority of studies on functionality have been concerned with their importance in bread and bread-type products. In the roller milling process one of the major objectives is to produce a white flour from the endosperm portion of the wheat kernel. This necessitates a repeated grinding and sieving operation to remove the outer components of the kernel, primarily the bran and germ. The chemical composition of the bran and endosperm portions of the wheat kernel varies and consequently usage of a wholewheat flour as opposed to a white flour will respond differently in end-product functionality. The major carbohydrate found in the endosperm or flour portion of the kernel is the starch. Other carbohydrate material includes free sugars, glucofructans and hemicelluoses (pentosans). In the outer layer of the wheat kernel (bran), the starch present is normally a contaminant from the endosperm. The carbohydrate materials found in this fraction would include cellulose, hemicellulose and free sugars.

This chapter will be concerned primarily with the carbohydrates found in the endosperm portion of the kernel and their importance as related to functionality of bread products. A broad overview of this subject is presented, with particular attention to certain studies conducted in the authors' laboratory.

8.2 Chemical composition and structure

8.2.1 Sugars and glucofructans

The sugars and glucofructans found in the endosperm of wheat are present as a minor component with values of 1–2% by weight having been reported. Individual sugars present include glucose, fructose, sucrose, and raffinose. Although certain reports have indicated the presence of maltose it is now

* Published with the approval of the Director, North Dakota Agricultural Experiment Station.

believed that the maltose was a degradation product of starch. Also present in the endosperm is a small amount of glucofructans which are water-soluble non-reducing oligosaccharides. The smallest member in this oligosaccharide series is sucrose followed by glucodifructose followed by sugars containing additional fructose units. The molecular weight of the glucofructans will extend in size to approximately 2000. Abou-Guendia and D'Appolonia[1] reported a progressive decrease in reducing and non-reducing sugars with wheat maturation. The sugar raffinose appeared at a particular stage of maturity of the wheat kernel and thereafter remained relatively constant.

8.2.2   Starch

The major carbohydrate component in wheat is the starch which provides an excellent source of energy. The unique physicochemical properties of starch impart particular properties to various types of foods. This aspect will be discussed later under functionality.

Starch represents 65–70% of wheat flour of 80% extraction or lower on a 14% moisture basis. Starch exists in the form of distinct granules and in the case of wheat is present in a bimodal distribution as illustrated in Figure 8.1.[2] Both large- and small-type granules are visible. It is these granules that are physically damaged during the roller milling operation and the conversion of the wheat into flour. The importance of such damage to overall bread functionality will be discussed later.

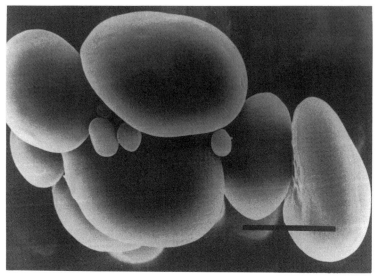

Figure 8.1 Bimodal distribution of wheat starch granules. Black bar is 10 μm long. From Fenno, ref. 2.

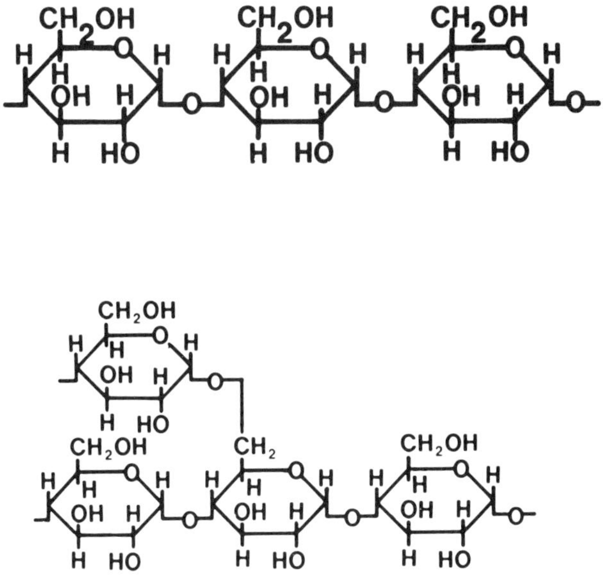

Figure 8.2 Structure of amylose and amylopectin.

Starch contains two major carbohydrate components, both high-molec-ular weight polymers. The amylose component is considered to be a linear polymer composed of glucopyranose units linked through α-D-$(1 \rightarrow 4)$ glycosidic linkages although there is now evidence to indicate that it is not completely linear. The second component, amylopectin, is a branched poly-mer with a reported molecular weight of 10^8 making it one of the largest found in nature. The basic structure for these two fractions is shown in Figure 8.2.

The mechanism by which the amylose and amylopectin components are synthesized, organized and incorporated into a granule is still not fully understood. Viewed under polarized light the starch granules have a posi-tive sign of birefringence. In addition, the granules produce distinctive X-ray powder diffraction patterns indicative, therefore, of areas of crystal-linity. Amylopectin is most probably the principal crystalline component of the granule. French[3] has reported on the layer or ring-like structure of the starch granule. The rings, often called growth rings represent concentric shells or layers of alternating high and low refractive index, density, crystal-linity, and resistance or susceptibility to chemical and enzymatic attack.

8.3 Gelatinization and pasting properties

Germane to any discussion concerning structure and functionality is the property of starch which causes it to undergo gelatinization and pasting when it is heated in aqueous solution. The changes that starch undergoes when it is heated with water are of paramount importance in terms of functionality in a wide array of food products. In the initial phase of this heating process, swelling begins in the least organized amorphous, inter-crystalline regions of the granule.[3] Further heating leads to uncoiling or dissociation of double helical regions and disappearance of the amylopectin crystallite structure. An increase in viscosity occurs. This phenomenon has been investigated by numerous researchers using the Brabender Amylo-graph instrument. Upon continued heating and hydration the granule will weaken to the point where it can no longer resist mechanical or thermal shearing, and a sol results.

Several reviews have been written on the gelatinization and pasting of wheat starch including that by Dengate.[4] The reader is referred to this report for a complete treatment of this area. There is further discussion on starch gelatinization and pasting as related to bread baking under the functionality section of this chapter.

8.4   Differential scanning calorimetry

The application of non-equilibrium theories of melting, annealing, and recrystallization of partially crystalline polymers and the concept of glass transition to the starch systems has resulted in development of a new research tool.[5,6] Differential scanning calorimetry (DSC) studies have been used to propose that the melting of the crystalline regions of starch are indirectly controlled by the continuous amorphous regions that surround them. Thus, a prerequisite of softening or plasticizing of the amorphous regions by the water present takes place before the non-equilibrium melting of the crystalline regions.[6] This plasticizing effect at the characteristic temperature, the glass transition temperature T_g, has favored sufficient mobility of the starch to be in a fluid or rubbery state called glass-to-rubber transition.[5,6]

The enthalpy and temperature of gelatinization of the endotherms obtained with DSC have been used to study the effect of processing and composition on the behavior of starch during gelatinization and retrogra-dation. These are important factors in the stability of most cereal-based products.[7]

Pure starch, amylose, and amylopectin models have been used to study the rate of recrystallization with an increasing content of retrograded B-type starch using DSC (Figure 8.3).[6,7] B-type starch is a crystalline hydrate that takes water from its environment, i.e. baked crumb. Retrograded amylo-pectin produces an endothermic curve at about 60°C, called a 'staling endotherm'. Retrograded amylose gels and the V-type crystalline amylose–lipid complex need higher temperatures (120–150°C) for their melting

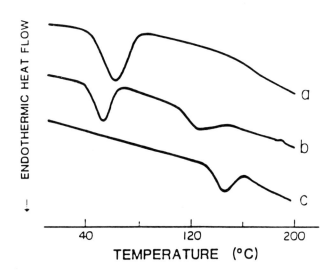

Figure 8.3 Typical DSC thermal curves of retrograded (a) waxy maize (40% w/w), (b) wheat starch (40% w/w), and (c) potato amylose (40% w/w) gels; following cooking, the gels were stored at 6°C for 6 h prior to calorimetry. From Biliaderis, ref. 7.

endotherm (Figure 8.3).[7] There are reports of quantitative DSC methods, using the area of the 'staling endotherm', to measure the rate of recrystallization and the effect of components and storage conditions in starch-based products.[6]

Several authors have reported an increase in gelatinization temperature, decreased gelatinization enthalpy, and extent of gelatinization by small molecular weight solutes, i.e. sugars and salts (Table 8.1). Competition for the available water, decrease in relative vapor pressure, sucrose inhibition of granular hydration, and specific sugar–starch interactions have been some of the proposed explanations.[6,8] No single parameter has interpreted completely the increase in gelatinization temperature caused by sugars. It has been suggested that a combination of parameters may predict contributions to free volume and local viscosity and thus help to explain this effect.[6] Lund[8] pointed out the need for more research in this area since a variety of starch-based products contain sucrose or salt in their formulation.

8.5   Nuclear magnetic resonance (NMR)

The use of multinuclear NMR to study the molecular mobility of starch and water have complemented DSC results on the study of gelatinization of

Table 8.1 Effect of salt and sucrose concentration on the gelatinization endotherm of wheat starch[a]

Conc. in aqueous phase (%)		ΔHg (cal g^{-1})	Extent of gelatinization (%)	Endotherm temperature		
				T_o	T_p	T_c
Salt	0	4.7	100	50	68	86
	3	2.7	57	58	71	88
	6	2.5	53	64	75	88
	9	2.6	55	68	78	88
	12	2.7	57	65	77	88
	15	2.7	57	65	77	88
	21	2.8	60	61	80	90
	30	3.3	70	59	79	91
Sucrose	15	3.2	68	50	70	86
	30	2.8	60	50	73	86
	45	2.3	49	50	75	86

[a] From Wootton and Bamunuarachchi 1980, as revised by Lund, ref. 8.

starch. Results showed that sucrose and NaCl reduced the mobility of D_2O, thus inhibiting the plasticizing effect of D_2O. During gelatinization an increased carbon chain mobility was observed as a drastic increase in ^{13}C signal intensity. Evidence was reported on the increase in 'trapped' water and on the sodium–starch interaction, and the decrease in ^{17}O and ^{23}Na relaxation time, respectively.[9]

8.6   Electron spin resonance (ESR) spectroscopy

ESR spectroscopy has been used to monitor the disappearance of the short-life and long-life free radicals formed during irradiation of starch and other foods.[10] Studies of irradiated pure starches showed that the initial ESR spectra are different for each starch type and may reflect differences in molecular arrangements of starch crystalline and amorphous fractions, i.e. distinct initial kinetics. The final ESR spectra seemed to be common for all the starches.[11]

8.7   Interactions

Wheat carbohydrates are a complex system that interacts with proteins, water, cosolutes, and lipids contributing to a range of functional properties. The study of structure and interactions of carbohydrates has used an array of techniques – including those mentioned – such as DSC, X-ray diffraction, NMR, ESR and several physical tests, i.e. rigidity of gels.[5]

8.7.1 Protein–starch interactions

Among the important factors affecting starch–protein interactions are the surface of the wheat starch granule and the nature of the gluten.[12,13] These interactions have also been studied as differences in endosperm texture, i.e. hardness vs. softness.[14–16] A 15 kDa polypeptide on the surface of starch granules has been hypothesized to interfere with the adhesion of the granules to the wheat protein matrix. This polypeptide has been reported to be absent in durum wheat, in small amounts in hard wheat, and present in soft wheat.[16] The technological importance of starch granule proteins, like the 15 kDa polypeptide, has been indirectly related to the chlorine gas treatment of flour and the effect on starch and the surface proteins.[16,17]

Adsorption of wheat proteins on starch granules revealed a greater affinity for the high-molecular weight proteins than for the low-molecular weight proteins.[18] The adsorption affinity of starch granules was higher after the granules were heated (40–80°C), at the pH range of 2.1–7.6, and in the presence of 0.0025 M NaCl.[18]

The influence of gluten addition on starch retrogradation, by measuring the change in rigidity and enthalpy during storage, showed a non-linear relationship between crystallinity and rigidity of retrograded starch gels (Figure 8.4).[19] As expected, the addition of gluten to starch gels caused a dilution of the gel structure yielding less rigid gels (Table 8.2). The competition for the water in the system and the dilution of the structural role of the starch were the proposed factors in the decreased retrogradation observed.[19]

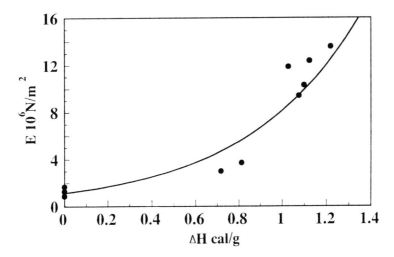

Figure 8.4 Variation of Young's modulus (E) with degree of crystallinity as determined by calorimetric methods (ΔH cal g^{-1}). Adapted from Forcinti and LeGrys, ref. 19.

Table 8.2 Shear stress at fracture for test starch gels[a]

	$10^5 \, N \, m^{-2}$
Starch gel	2.57
6% Added gluten	2.34
14% Added gluten	1.99
14% Added gluten (developed)	1.70

[a] From Forcinti and Le Grys, ref. 19.

8.7.2 Lipid–starch interactions

An increasing number of reports on the study of starch–lipid interactions using stable free radical spin probes in ESR spectroscopy can be found in the literature. ESR studies showed that the probes 2,2,6,6-tetramethyl piperidine-1-oxyl (TEMPO)-laurate and 16-Doxyl stearic acid were bound to starch granules in the presence of water.[20] The study showed that the lipid probe TEMPO-laurate in water was equally bound to amylose and amylopectin. However, the same probe caused granule disintegration and loss of birefringence in waxy corn.[20]

The interaction of lipids with proteins and carbohydrates during baking have been reviewed by Pomeranz and Chung.[21] In dough systems glycolipids were reported to interact mainly with gluten, while in baked bread systems, glycolipids interacted mainly with starch.[22]

An amylopectin–lipid transition above 100°C on DSC thermograms has been reported.[23] Evidence of the formation of amylopectin–surfactant/emulsifiers (sodium dodecyl sulfate and cetyltrimethylammonium bromide) was presented as X-ray diffraction components of B- and V-patterns.[23] Slade and Levine[24] also reported a thermal transition of waxy maize starch–sodium stearoyl lactylate (10 : 1 w/w). Amylopectin formed a complex with a thermal transition at about 70°C. The delay in starch retrogradation in breads and lean baked products during storage has been explained by the formation of micelles with a V-type X-ray diffraction pattern of the amylose fraction. However, recent evidence suggests that amylopectin outer chains may also form this V-type complex.[23,24]

Starch–lipid interactions have also been studied following the endotherm curves of DSC in the presence of cetyltrimethylammonium bromide (CTAB).[25] As expected the addition of the ligand yielded an amylose–lipid transition enthalpy, but it also lowered the gelatinization enthalpy.[25]

8.7.3 Hemicelluloses (pentosans)

Pentosans represent a minor component of the wheat endosperm (2–3%) yet they have been studied by many researchers in terms of end-product and their specific effect on dough and bread properties.

A portion of the pentosans in the endosperm or flour portion of the kernel

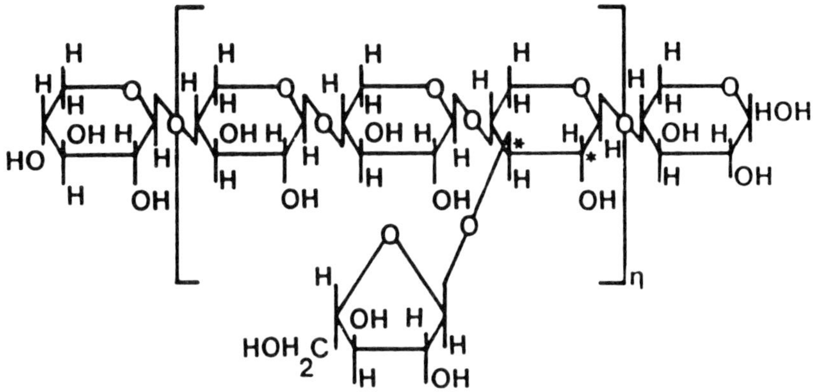

Figure 8.5 Structure of wheat endosperm pentosans; possible arabinose branching points at C2 and C3 are designated by *.

is readily extractable by water. The remaining portion is extracted by alkali from a layer of material located directly on top of the prime starch after centrifugation of an aqueous slurry of flour and water. This layer of material has been given different names by different workers including 'sludge', 'squeegee' and 'amylodextrin'. The pentosans found in this fraction are water soluble once extracted by alkali.

Pentosan preparations obtained by water extraction normally contain considerable amounts of protein which require purification to obtain an essentially pure arabinoxylan. The major sugars found in the arabinoxylan are xylose and arabinose. Xylose forms the main backbone chain while arabinose occurs as part of the branching. The structure of the wheat endosperm pentosans (arabinoxylan) is given in Figure 8.5

Two techniques that have been used in our laboratory for the isolation and purification of the water-soluble and water-insoluble pentosans from wheat flour are illustrated in Figures 8.6 and 8.7, respectively.

Evidence also exists for the presence of an arabinogalactan in the wheat endosperm. Since pentosans are essentially 'gums' they do have a very high water-binding capacity. They have also been implicated in the oxidative gelation reaction of flour extracts caused by the addition of hydrogen peroxide. These two phenomena will be discussed in greater detail in the discussion on functionality.

8.8 Functionality as related to bread products

8.8.1 Sugars and glucofructans

The small amount of sugars and glucofructans present in the wheat endosperm do not play a significant part in rheological properties. The main

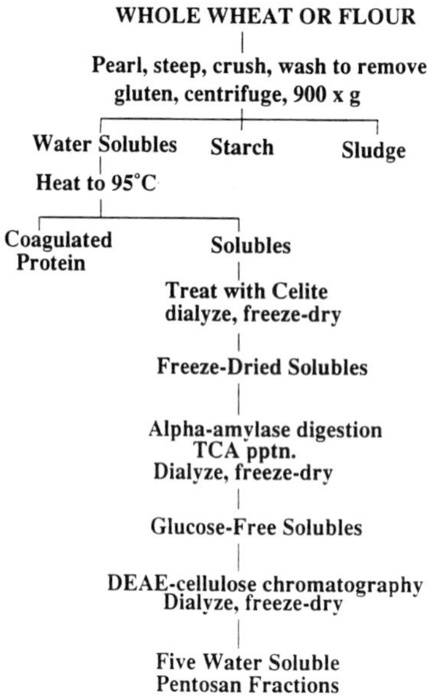

WHOLE WHEAT OR FLOUR
|
Pearl, steep, crush, wash to remove
gluten, centrifuge, 900 x g

Water Solubles   Starch   Sludge
|
Heat to 95°C

Coagulated   Solubles
Protein
|
Treat with Celite
dialyze, freeze-dry
|
Freeze-Dried Solubles
|
Alpha-amylase digestion
TCA pptn.
Dialyze, freeze-dry
|
Glucose-Free Solubles
|
DEAE-cellulose chromatography
Dialyze, freeze-dry
|
Five Water Soluble
Pentosan Fractions

Figure 8.6 Isolation and purification procedure for water-soluble pentosans.

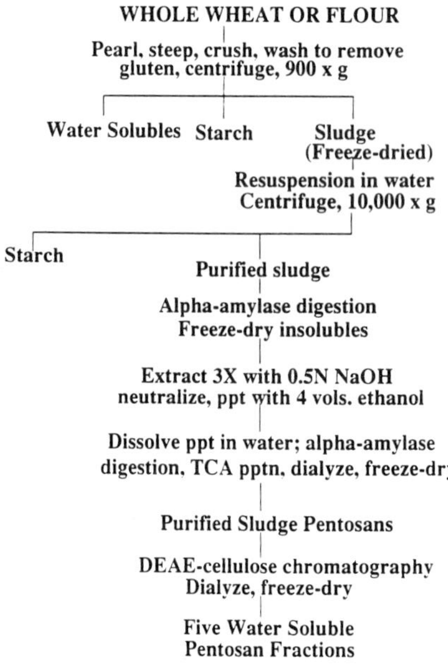

WHOLE WHEAT OR FLOUR
|
Pearl, steep, crush, wash to remove
gluten, centrifuge, 900 x g

Water Solubles   Starch   Sludge
(Freeze-dried)
|
Resuspension in water
Centrifuge, 10,000 x g

Starch   Purified sludge
|
Alpha-amylase digestion
Freeze-dry insolubles
|
Extract 3X with 0.5N NaOH
neutralize, ppt with 4 vols. ethanol
|
Dissolve ppt in water; alpha-amylase
digestion, TCA pptn, dialyze, freeze-dry
|
Purified Sludge Pentosans
|
DEAE-cellulose chromatography
Dialyze, freeze-dry
|
Five Water Soluble
Pentosan Fractions

Figure 8.7 Isolation and purification procedure for water-insoluble pentosans.

Table 8.3 Change in glucose and fructose content during fermentation

Stage	Fructose (%)	Glucose (%)
After mixing	3.22	3.07
1 h fermentation	3.12	2.56
2 h fermentation	3.08	2.12
3 h fermentation	2.31	1.06
After proofing	1.65	0.63
Bread crumb	1.28	0.40

function of these sugars is to act as a substrate for the yeast to act upon during fermentation. Without the incorporation of additional sugar to the bread formulation these sugars would not play a role in the browning reaction during the baking stage since they would be used up during fermentation. Any residual or added sugars would be involved in the caramelization and Maillard reactions, thereby contributing to the brown color of the crust.

It has been reported by several workers that sucrose added to a baking formula is hydrolyzed almost immediately during the mixing stage to glucose and fructose. Work in this laboratory has shown that glucose is preferentially fermented over fructose when sucrose is added to the formula as shown in Table 8.3. In a bread formulation with no added sugar the naturally occurring fructose and glucose are fermented immediately and thereafter the yeast will adapt to maltose.

8.8.2 Starch

Many reviews have been written on the role of starch on dough and bread functionality. Although the quantity and quality of the protein in wheat are considered major factors of importance in determining a high quality bread wheat the carbohydrate components including the starch also play an important role. It is generally recognized that wheat starch compared to other starches is unique in terms of bread functionality. Hoseney et al.[26] found that only wheat, barley, and rye starches produce satisfactory bread, because they have very similar properties, including gelatinization temperature range, granule shape, and bimodal starch granule size distribution as illustrated in Table 8.4. The normal amylose : amylopectin ratio, found in the wheat starch granule, is required for satisfactory bread performance.

Although more than 30 years have now elapsed since Sandstedt[12] outlined five functions for starch in baking, their importance is still applicable today as we discuss starch functionality (Table 8.5). Starch does play a role in diluting the gluten to a desirable consistency. It is impossible to conceive of a loaf of bread as we know it made from 100% gluten. The role of damaged starch producing fermentable sugars by the action of the amylase enzymes

Table 8.4 Baking data for reconstituted flours containing various starches[a]

	Mixing time (min)	Absorption (%)	Loaf volume (cm³)
Wheat	2.8	61	80
Barley	2.5	61	78
Rye	3.1	61	77
Corn	3.8	75	48
Oat	2.6	70	58
Rice	1.8	73	68

[a] From Hoseney *et al.*, ref. 26.

during the fermentation stage of baking is well recognized. It is desirable to have a certain amount of starch damage in the flour to obtain a satisfactory loaf of bread. Excessive amounts of starch damage, however, will result in deterioration of bread quality. It is during the mixing stage that the starch granules become embedded in the gluten matrix and also during mixing that the gluten and starch become bound together by relatively strong electrostatic forces.[27] The surface characteristics of the starch granules affect starch–protein interactions. Kulp and Lorenz[13] found adhering matter on starch granules obtained from both wheat and flour. These workers concluded that the interaction among the gluten and starch components of flour affects mixing and dough characteristics.

The pasting or gelatinization properties of starch are important in terms of its functionality in bread products. It is during the baking stage that limited gelatinization occurs in the bread as the result of heat application. Since there is limited water available not all of the starch will completely gelatinize. Figure 8.8 shows the presence of gelatinized wheat starch granules in bread crumb. It is also during this stage that amylase action on the starch will occur and that, if excess amounts of amylase enzyme are present, the bread crumb will become very sticky. It is important to

Table 8.5 Functions of starch in baking[a]

1. Dilutes gluten to desirable consistency
2. Furnishes sugar through amylase action
3. Furnishes surface suitable for strong union with gluten
4. Becomes flexible but does not disintegrate during partial gelatinization
5. Sets structure to the final loaf of bread

[a] From Sandstedt, ref. 12.

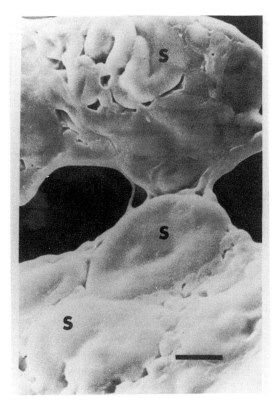

Figure 8.8 Gelatinized wheat starch granules (denoted by 'S') in bread crumb. Black bar is 10 μm long. From Fenno, ref. 2.

recognize that since cereal-based products can vary in the amount of available water, the function of the starch will also differ. For example, in a cake system where there is considerably more water present, the importance of starch may differ from a bread product where water is limiting. The restricted type of starch gelatinization that occurs in bread, coupled with the gluten denaturation, contribute in setting up the structure of the final loaf of bread.

Although many studies have been undertaken to determine the importance of starch in baking, more research is still required to elucidate fully the exact role of this biochemical component.

A major role for starch in bread occurs once the bread has been removed from the oven and the process of staling or aging begins. The entire subject of bread staling has been researched for many years and numerous reviews have been published on the topic. It is not within the scope of this chapter to discuss the topic in any significant detail.

Table 8.6 Methods used to measure staling

1. Firmness techniques
2. Taste panel evaluations
3. Changes in starch
   a. Decrease in soluble starch
   b. Decrease in enzyme susceptibility of starch
   c. Increase in starch crystallinity
   d. Changes in X-ray diffraction patterns
   e. Changes in differential scanning calorimetry patterns

Although factors other than starch may contribute to the staling process, including ingredients, baking procedures, and other biochemical components, the importance of the starch is well documented. Bread staling has been defined by various workers as 'decreasing consumer acceptance of bakery products caused by changes in the crumb other than those resulting from the action of spoilage organisms'. The terms staling and retrogradation are often used interchangeably, however, the distinction must be made that staling refers to all of the changes that take place in a loaf of bread, whereas, retrogradation refers specifically to changes in the starch components. Since starch is believed to be the major component responsible for staling, many of the methods used to measure this phenomenon are dependent on changes in

Table 8.7 Effect of staling on the quantity and composition of soluble starch extracted from bread crumb stored at 21°C[a,c]

Bread[b]	Day	Soluble starch (%)	Composition of soluble starch	
			Amylose (%)	Amylopectin (%)
A	0	3.34	0.52	2.82
	1	2.16	0.19	1.97
	2	1.76	0.14	1.58
	5	1.22	0.10	1.12
B	0	2.36	0.39	1.97
	1	1.60	0.12	1.48
	2	1.14	0.08	1.08
	5	1.08	0.07	1.02
C	0	1.49	0.06	1.43
	1	1.35	0.04	1.31
	2	1.12	0.04	1.09
	5	1.02	0.03	0.99

[a] All results reported on a dry basis. [b] A, B, and C flour protein 11, 13.9, and 21.9%, respectively on 14% mb. [c] From Kim and D'Appolonia, refs. 28 and 30.

Table 8.8 Effect of staling on the quantity and composition of soluble starch extracted from bread crumb B during 12 h storage[a,d]

Time (h)	Storage temperature (°C)	Soluble starch (%)	Composition of soluble starch	
			Amylose	Amylopectin[b]
0.1	Room Temp.	2.51	0.60	1.91
2	Room Temp.	2.34	0.39	1.95
5[c]	21	1.86	0.22	1.64
	30	1.92	0.25	1.67
12[c]	21	1.74	0.18	1.56
	30	1.85	0.21	1.64

[a] All results reported on a dry basis. [b] By difference. [c] Breads were cooled for 2 h at room temperature and then stored at 21 and 30°C for 3 and 10 h, respectively. [d] From Kim and D'Appolonia, refs. 28 and 30..

the starch. Some of the methods used to measure staling are listed in Table 8.6. The effect of staling on the quantity and composition of soluble starch extracted from bread crumb produced from flour of different protein content was studied by Kim and D'Appolonia.[28] These workers showed (Table 8.7) that the recovery of the soluble starch from fresh bread crumb is inversely related to the protein content of the flour. Although the amount of amylose leached from fresh bread crumb was small it decreased during bread staling. The largest decrease, however, occurred during the first day of storage; thereafter the changes were small. Data on the recovery of soluble starch and its composition extracted from bread crumb B during the first 12 h of storage after baking are presented in Table 8.8. A sharp decline in amylose content occurred during the first few hours of storage. Most of the amylose retrogradation takes place during baking and subsequent cooling of the loaf. In studies conducted in this laboratory using different storage temperatures it has been shown that at elevated temperatures some factor (changes in protein or moisture redistribution or both) in addition to starch, plays an important role in the firming process undergone by bread. At room temperature, staling is characterized primarily by starch crystallization.

8.8.3 Pentosans

As indicated previously the pentosans are a minor component of wheat flour representing 2–3%. Numerous researchers have studied their importance in terms of bread functionality and conflicts do appear in the literature relative to their importance. Various techniques have been used to study the effect of this biochemical component on bread properties. Several studies have involved extraction and purification followed by incorporation of the

Table 8.9 Effect of water solubles, pentosans and fractionated pentosans extracted from HRS (Thatcher) flour on gluten–starch loaves[c]

Addition[a]	Loaf volume		Crust color[b]	
	I (cm³)	II (cm³)	I	II
Control (gluten–starch)	127	132	2	2
3.2% Water solubles	191	188	3	3
0.8% Water solubles	138	155	2+	3
0.8% Crude pentosans	142	147	2–	2–
0.8% Amylase-treated pentosans	146	145	2–	2–
0.8% F1 + F2A + F2B	135	135	1+	1+
0.8% F3 + F4	147	144	2+	2
0.8% F4 + F5	153	151	3	2+

[a] Percentage of additive based on 100 g flour. [b] Based on a 1–4 scale, 4 = very good and 1 = poor. [c] From D'Appolonia *et al.*, ref. 29.

material to a base flour or to a gluten–starch system and the effect on baking quality observed. In other studies specific enzymes with high pentosanase activity have been added to the baking formula to degrade the pentosans and the resulting effect on bread volume and characteristics noted. Since pentosans are gums, they do have a high water-binding capacity and consequently it is well established that they have a positive increasing effect on absorption when added to a dough system. The extraction of pentosans from the wheat endosperm or wheat flour results in a material which contains considerable amounts of protein. If such material is added to a wheat flour or gluten–starch system a positive increase in loaf volume is noted. This positive effect may be due to the adhering protein or to a protein carbohydrate complex. Table 8.9 shows results obtained from a study in this laboratory[29] on the effect of pentosans on bread properties. Addition of pure pentosan material (F1 + F2), basically arabinoxylan, resulted in only a negligible increase in loaf volume, but a detrimental effect on crust color was

Table 8.10 Effect of pentosans on the Avrami exponent and the time constant of bread stored at 21°C[a]

Bread	Avrami exponent	Time constant	Time constant over first day of storage
Control	0.92	5.44	4.80
With 1.0% soluble pentosans	0.73	6.53	4.23
With 1.0% insoluble pentosans	0.77	8.54	5.88

[a] From Kim and D'Appolonia, ref. 30.

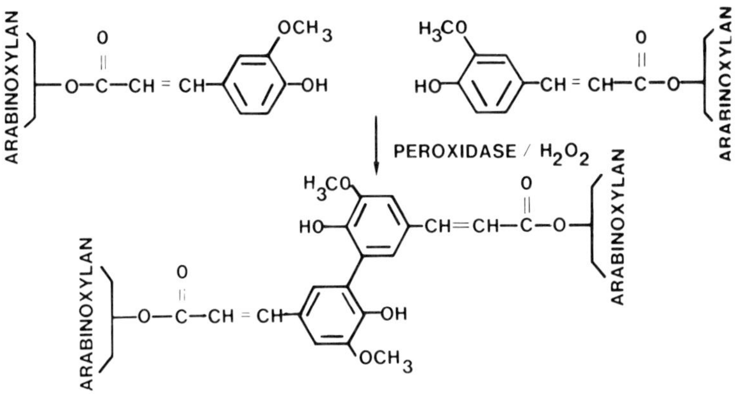

Figure 8.9 Suggested mechanism for the oxidative phenolic coupling of two ferulic acid residues. From Neukom and Markwalder, ref. 31.

observed. With addition of fractions F3, F4, and F5, which also contained protein material, a greater effect was noted on loaf volume.

In addition to the effect of pentosans on dough and bread properties they have also been implicated in bread staling. The incorporation of 1% soluble pentosans and 1% insoluble pentosans in bread and the effect on staling was studied in this laboratory.[30] The effects of the incorporation of this pentosan material on the Avrami exponent and time constant are presented in Table 8.10. The data indicate that pentosans decreased the bread staling rate, and the effect exerted by the water-insoluble pentosans was more pronounced than that exerted by the water-soluble pentosans. Kinetic studies indicated that pentosans simply reduce the amount of starch components available for crystallization, thus decreasing the bread staling rate.

Pentosans have been implicated in the gelation reaction which occurs when a water extract from wheat flour is treated with trace amounts of an oxidizing agent. Ferulic acid, which is also involved in this reaction, is believed to be associated with the pentosan component. A mechanism for this oxidative gelation reaction has been proposed by Neukom and Markwalder[31] (Figure 8.9).

Upon oxidation a dimer of ferulic acid is formed. In this laboratory we were interested to determine if this reaction occurs with all classes of wheat and with different mill streams. Utilizing five classes of wheat Ciacco and D'Appolonia[32] reported differences in viscosity with time upon addition of an oxidizing agent (H_2O_2/peroxidase) to arabinoxylan extracted from the different wheat classes. The highest intrinsic viscosity values were obtained for the arabinoxylan isolated from western white wheat, whereas that from soft red winter wheat had the lowest values.

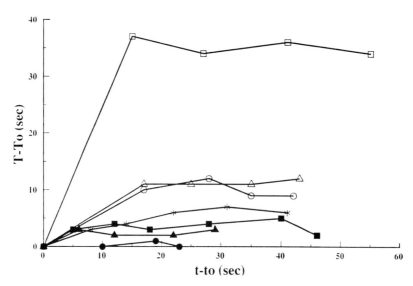

Figure 8.10 Increase in viscosity with time upon addition of H_2O_2/peroxidase to solutions of F_1 (arabinoxylan fraction). $T–T_o$ = flow, $t–t_o$ = gelation time. □, 6 Midds; △, 4 break; ○, tailings; *, 1 break; ■, low quality; ▲, 5 break; ●, 1 midds. From Ciacco and D'Appolonia, ref. 33.

The same workers in another study[33] showed (Figure 8.10) that the arabinoxylans isolated from flour streams representing primarily the inner portion of the kernel had higher intrinsic viscosity, were less branched, and in general had higher gelling capacity than those isolated from flour streams containing a greater percentage of the outer portion of the kernel. The importance of the gelation reaction in terms of the baking process itself is unclear.

8.9   Effects of gamma irradiation on processing

The use of gamma irradiation processing in foods ranges from low doses to inhibit sprouting, ripening, and to promote insect disinfestation (up to 1 kGy), to high doses to sterilize and eliminate viruses (10–70 kGy).[34] In synthesis, the main effects of irradiation treatment on wheat carbohydrates are at the molecular level, hydrolyzing covalent bonds. At high doses starch is hydrolyzed giving lower molecular weight amylose and amylopectin.[10,35,36] Reduction of viscosity, swelling power, and iodine-binding power are also reported. These changes are accompanied by increased water solubility, acidity, and reducing sugars.[10,34] Irradiation has been reported to increase the amount of soluble pentosans, but the total quantity of pentosans was unchanged.[34] In barley, an increase in soluble β-glucans was also reported.[36]

Studies on gamma radiation of wheat have been conducted in our laboratory. In one study the effects of low-dosage radiation on starch properties were studied and in the other the effects on dough and baking properties were examined.[37,38]

Upon gamma-irradiation treatments of wheat (50, 100, 200, and 300 krad) the isolated starch showed a decrease in peak viscosity, increased water-binding capacity, decreased swelling power, and increased solubility.[38] As irradiation dosage was increased in the wheat samples the resultant flour gave an increase in farinograph absorption and dough development time while stability decreased. Baking quality decreased as radiation increased but appeared to be influenced by variety.

8.10 Summary

This chapter has attempted to give an overview of the structure and function of wheat carbohydrates with particular emphasis on certain studies undertaken in the Department of Cereal Science at North Dakota State University.

References

1. Abou-Guendia, M. and D'Appolonia, B. L. (1972) Changes in carbohydrate components during wheat maturation. I. Changes in free sugars. *Cereal Chem.*, **49**, 664–676.
2. Fenno, C. W. (1984) Use of specially treated starches in bread baking. *PhD Thesis*, North Dakota State University, Fargo.
3. French, D. (1984) Organization of starch granules, in *Starch: Chemistry and Technology*, 2nd edn, (ed. R. L. Whistler, J. N. BeMiller and E. F. Paschall). Academic Press, New York, pp. 183–247.
4. Dengate, H. N. (1984) Swelling, pasting and gelling of wheat starch, in *Advances in Cereal Science and Technology*, Vol. VI, (ed. Y. Pomeranz). Am. Assoc. Cereal Chem., St. Paul, Minnesota, pp. 49–82.
5. Morris, V. J. (1990) Starch gelation and retrogradation. *Trends Food Sci. Technol.*, 2–6.
6. Slade, L. and Levine, H. (1989) A food polymer science approach to selected aspects of starch gelatinization and retrogradation, in *Frontiers in Carbohydrate Research – 1 Food Application*, (ed. R. P. Millane, J. N. BeMiller and R. Chandrasekaran). Elsevier Applied Science, London, pp. 215–270.
7. Biliaderis, C. G. (1991) The structure and interactions of starch with food constituents. *Can. J. Physiol. Pharmacol.*, **69**, 60–78.
8. Lund, D. (1984) Influence of time, temperature, moisture, ingredients, and processing conditions on starch gelatinization. *CRC Critical Rev. Food Sci. Nutrit.*, **20**, 249–273.
9. Chinachoti, P., White, V. A., Lo, L. and Stengle, T. R. (1991) Application of high-resolution carbon-13, oxygen-17 and sodium-23 nuclear magnetic resonance to study the influences of water, sucrose and sodium chloride on starch gelatinization. *Cereal Chem.*, **68**, 238–244.
10. Rayas-Duarte, P. and Rupnow, J. R. (1989) Some physical and chemical properties of gamma irradiated food starches, in *Frontiers in Carbohydrate Research – 1 Food Applications*, (ed. R. P. Millane, J. N. BeMiller and R. Chandrasekaran). Elsevier Applied Science, London, pp. 171–199.
11. Raffi, J., Angel, J. P., Thiery, C. J., Frejville, C. M. and Saint-Lebe, L. (1981) Study of

gamma irradiated starches derived from different foodstuffs. A way for extrapolating wholesomeness data. *J. Agric. Food Chem.*, **29**, 1227–1232.

12. Sandstedt, R. M. (1961) The function of starch in the baking of bread. *Baker's Digest*, **35**(3), 36–43.

13. Kulp, K. and Lorenz, K. (1981) Starch functionality in white pan breads: new developments. *Baker's Digest*, **55**(5), 24–28, 36.

14. Barlow, K. K., Buttrose, M. S., Simmonds, D. H. and Vesk, M. (1973) The nature of the starch–protein interface in wheat endosperm. *Cereal Chem.*, **50**, 443–454.

15. Simmonds, D. H., Barlow, K. K. and Wrigley, C. W. (1973) The biochemical basis of grain hardness in wheat. *Cereal Chem.*, **59**, 553–562.

16. Schofield, J. D. and Greenwell, P. (1986) Wheat starch granule proteins and their technological significance, in *Cereals in a European Context*. First European conference on food science and technology, (ed. I. D. Morton and E. Horwood). London, pp. 407–420.

17. Gough, B. M., Greenwood, C. T. and Whitehouse, M. E. (1978) The role and function of chlorine in the preparation of high ratio-cake flour. *CRC Crit. Rev. Food Sci. Nutrit.*, **10**, 91–113.

18. Eliasson, A. and Tjerneld, E. (1990) Adsorption of wheat proteins on wheat starch granules. *Cereal Chem.*, **67**, 366–372.

19. Forcinti, R. and LeGrys, G. A. (1989) The effect of gluten content on the staling of wheat starch, in *Trends in Food Science*, (ed. A. H. Ghee, L. W. Sze and F. C. Woo). Proceedings of the 7th World Congress of Food Science and Technology, Singapore, pp. 179–181.

20. Nolan, N. L., Faubion, J. M. and Hoseney, R. C. (1986) An electron spin resonance study of native and gelatinized starch systems. *Cereal Chem.*, **63**, 287–291.

21. Pomeranz, Y. and Chung, O. K. (1978) Interaction of lipids with proteins and carbohydrates in breadmaking. *J. Am. Oil Chem. Soc.*, **55**, 285–289.

22. Pomeranz, Y. (1973) Interactions between glycolipids and wheat flour macromolecules in breadmaking. *Adv. Food Res.*, **20**, 153–188.

23. Gudmundsson, M. and Eliasson, A. C. (1990) Retrogradation of amylopectin and the effects of amylose and added surfactants/emulsifiers. *Carbohydrate Polym.*, **13**, 295–315.

24. Slade, L. and Levine, H. (1987) Recent advances in starch retrogradation, in *Recent Developments in Industrial Polysaccharides*, (ed. S. S. Stivala, V. Crescenzi and I. C. M. Dea). Gordon and Breach Science, New York, pp. 287–430.

25. Eliasson, A., Finstad, H. and Ljunger, G. (1988) A study of starch-lipid interactions for some native and modified maize starches. *Starch*, **40**, 95–100.

26. Hoseney, R. C., Finney, K. F., Pomeranz, Y. and Shogren, M. D. (1971) Functional (breadmaking) and biochemical properties of wheat flour components. VIII. Starch. *Cereal Chem.*, **48**, 191–201.

27. Medcalf, D. G. and Gilles, K. A. (1968) The function of starch in dough. *Cereal Sci. Today*, **13**(10), 382–385, 388, 392–393.

28. Kim, S. K. and D'Appolonia, B. L. (1977) Bread staling studies. II. Effect of protein content and storage temperature on the role of starch. *Cereal Chem.*, **54**, 216–224.

29. D'Appolonia, B. L., Gilles, K. A. and Medcalf, D. G. (1970) Effect of water-soluble pentosans on gluten-starch loaves. *Cereal Chem.*, **47**, 194–204.

30. Kim, S. K. and D'Appolonia, B. L. (1977) Bread staling studies. III. Effect of pentosans on dough, bread, and bread staling rate. *Cereal Chem.*, **54**, 225–229.

31. Neukom, H. and Markwalder, H. U. (1978) Oxidative gelation of wheat flour pentosans: A new way of cross linking polymers. *Cereal Foods World*, **23**, 374–376.

32. Ciacco, C. F. and D'Appolonia, B. L. (1982) Characterization of pentosans from different wheat flour classes and of their gelling capacity. *Cereal Chem.*, **59**, 96–100.

33. Ciacco, C. F. and D'Appolonia, B. L. (1982) Characterization and gelling capacity of water-soluble pentosans isolated from different mill streams. *Cereal Chem.*, **59**, 163–166.

34. Urbain, W. M. (1986) Radiation chemistry of food components and of foods, in *Food Irradiation*, (ed. W. M. Urbain). Academic Press, Orlando, pp. 37–115.

35. Roushdi, M., Harras, A., El-Meligi, A. and Bassim, M. (1983) Effect of high doses of gamma rays on corn grains. Part II. Influence on some physical and chemical properties of starch and its fractions. *Starch*, **35**, 15–19.

36. Bhatty, R. S. and MacGregor, A. W. (1988) Gamma irradiation of hull-less barley: Effect on grain composition, β-glucans and starch. *Cereal Chem.*, **65**, 463–470.

37. MacArthur, L. A. and D'Appolonia, B. L. (1983) Gamma radiation of wheat. I. Effects on dough and baking properties. *Cereal Chem.*, **60**, 456–460.
38. MacArthur, L. A. and D'Appolonia, B. L. (1984) Gamma radiation of wheat. II. Effects of low-dosage radiations on starch properties. *Cereal Chem.*, **61**, 321–326.

9 Wheat lipids: structure and functionality

W. R. MORRISON

9.1 Composition and distribution

9.1.1 Composition

The majority of the lipids in wheat are fatty acid (FA) esters of glycerol, and the remainder include free (unesterified) fatty acids (FFA) and several types of sterol-based lipids and glycosphingolipids.[1,2] The major glycerolipids are triglyceride (TG), mono- and di-galactosyldiglycerides (MGDG, DGDG), N-acylphosphatidylethanolamine (NAPE), phosphatidylethanolamine (PE), phosphatidylglycerol (PG) and phosphatidylcholine (PC). Small amounts of the partially acylated glycerolipids are also found – these include diglyceride (DG), the *mono* (or *lyso*)acyl glycerides *MG*, M*GMG*, D*GMG*, NA*L*PE, *L*PE, *L*PG and *L*PC.

For all practical purposes the non-polar or neutral lipids (NL), comprised of TG, DG, MG and the non-polar sterol lipids, have similar properties and behave as a single entity. The same applies to the glycolipids (GL), comprised of DGDG, MGDG, DGMG, MGMG and several other minor components. Sometimes the phosphoglycerides or phospholipids (PL) can be treated as a single group, but for most purposes the starch lysoPL (LPL) should be distinguished separately. The total polar lipids (PoL) are comprised of the GL and PL (including LPL).

The principal FAs in the glycerolipids are palmitic (16:0), stearic (18:0), oleic (18:1), linoleic (18:2) and linolenic (18:3), and the composition of the major lipid groups is highly unsaturated (Table 9.1). From the nutritional point of view there is a high proportion of the n-6 type of polyunsaturated acid (18:2) and also some of the important n-3 type (18:3). In wheat technology 18:2 and 18:3 are susceptible to non-enzymic and enzymic oxidations (see section 9.2.2).

The tocopherols also deserve some mention since they are important antioxidants and have a high vitamin E value. In wheat they are comprised almost exclusively of α- and β-tocols and the corresponding tocotrienols.[1,2,4] Much may be destroyed on baking.

9.1.2 Distribution in the caryopsis

The lipid reserves in wheat are stored in oil droplets or spherosomes located mostly in the scutellum and aleurone, but also in the subaleurone starchy

Table 9.1 Fatty acid composition of some lipid fractions in wheat[a]

Lipid source	Fatty acid (wt. %)				
	16:0	18:0	18:1	18:2	18:3
Whole grain, total	17–24	1–2	18–21	55–60	3–5
Germ, total	18–24	2	8–17	54–57	4–9
Endosperm, total	18	1	19	56	3
Flour					
Non-starch, total	16–21	<2	12–13	60–66	4–5
Non-starch NL	17–19	1–2	14–15	60–62	4–5
Non-starch GL	12–14	1–2	8–9	71–74	4–5
Non-starch PL	23–27	1	9–10	59–63	2–4
Starch lipid	35–44	<2	1–14	44–52	1–4

[a] From Morrison, refs. 1–3.

endosperm.[5] Spherosomes are the source of the NL and tocopherols in wheat, but they include some of the PL in their boundary membrane.

The diacyl polar lipids are probably all structural components of various membranes, and there is evidence that almost all of the GL, NAPE and NALPE are from the amyloplast membranes within which the starch granules were formed. The structural lipids in other organelles and membranes found in all tissues are the ubiquitous PL, i.e. PC, PE and some other minor PL.[2] The starch granules contain LPC, LPE and LPG;[3] traces of FFA may be adsorbed artefacts from the other endosperm lipids.[6] It is sometimes useful to distinguish starch lipids, which are comparatively inert in wheat, flour and dough (before baking), from the other non-starch lipids (NSL) which participate in several chemical and physical events (see section 9.2).

The distribution of lipids in the wheat caryopsis is summarized in Table 9.2. Storage lipids (TG and other NL) are concentrated in the scutellum and aleurone, while the GL and N-acylPL are almost exclusively in the starchy endosperm. The majority of the LPL are in the starch granules, but some LPL are formed by partial hydrolysis of diacyl-PL in the NSL and they are not readily distinguishable. Pericarp tissue contains only remnants of lipids since they are extensively degraded during the senescence that accompanies grain ripening.

9.1.3 Varietal differences and edaphic effects

Differences in lipid content among wheats are quite common, but the composition of the NL and PoL lipid classes and the FA composition of each lipid remain fairly constant. Most variation can be attributed to differences in the proportions of the major tissues in large and small grains (and hence to the balance of reserve lipids to structural lipids) plus the extent to which

Table 9.2 Distribution of lipids in the wheat grain[a]

Grain fraction	Weight of fraction[b]	Lipid groups[c]			Total lipid	
		NL	GL	PL	Tissue[d]	Grain[e]
Whole grain	100	44–57	8–14	31–42	2.5– 3.3	100
Germ	2.5– 3.0	79–85	<4	13–17	25.7–30.4	25–30
Aleurone	4.0–10.0	72–83	2–10	14–18	8.6–10.6	22–23
Pericarp	6.8– 8.6	6	2	<1	1.3	4
Endosperm (for NSL)	78.7–84.5	33–47	20–35	22–35	0.7– 1.1	20–31
Starch in endosperm	59.3–67.5	4– 6	1– 7	89–94	0.7– 1.2	16–22

[a] Calculated from data given by Hargin and Morrison, ref. 7; and Morrison, ref. 2. [b] Weight percentage of whole grain (dry basis). [c] Weight percentage of total lipid in tissue. [d] Weight percentage of lipid in tissue (dry basis). [e] Weight percentage of total lipid in grain.

edaphic factors favoured the synthesis of storage lipids during the grain-filling period. Thus, characteristic differences in NSL are to be expected between spring- and winter-types of wheat, and there can be appreciable effects from the environment in those areas where the weather is less stable during the grain filling period.

Ambient temperature during grain filling affects the lipid content of starch granules in wheat and barley.[8–10] Since starch lipids inhibit the extent to which starch granules swell during cold or hot gelatinization (see section 9.2.6), this could contribute to longer term variation in the baking quality of bread and cake flours.

9.1.4   Redistribution of lipids in millstreams

During milling some lipids are transferred from the germ and aleurone to the endosperm particles which are eventually reduced to flour.[5,7,11,12] Since the tocopherols in the germ (scutellum) are exclusively α- and β-tocols, while those in the aleurone and starchy endosperm are the corresponding tocotrienols, they provide convenient markers to follow the redistribution of lipids from these tissues during milling.[13,14] Simultaneous transfer of ash and protein, which are clearly non-lipid components, indicates that particles of essentially intact tissue are also transferred. The breakdown of tissue and the amounts of lipid transferred depend on milling practice, including the moisture content to which the grain is conditioned.[14,15]

There is a simple model which explains the composition of the lipids in millstreams.[14] Each millstream has a base level of NL, GL and PL, derived from the starchy endosperm, which is fairly constant in composition. To this is added a highly variable amount of lipid (together with ash, protein and material contributing to colour) from germ and aleurone, roughly in the proportion 3:1. The transferred lipid is NL with about 10–15% PL but no GL, and it is very similar in composition to wheat germ oil.

9.2 Functionality

9.2.1 Lipolytic enzymes

Lipids are comparatively stable in the whole caryopsis, and changes occur very slowly in sound stored grain with only a gradual loss of baking quality over several years.[16,17] However, deterioration in baking quality is much faster in damp stored flour, and is associated with the proliferation of moulds and extensive hydrolysis of NSL to FFA.[18,19] Control of storage atmosphere can be used to retard lipid hydrolysis,[20,21] as well as employing the standard practice of controlling moisture content and temperature.

Changes in lipids occur much faster in milled products due to the dispersal of lipid and improved access of enzymes to substrates. Hydrolysis is particularly rapid in wholemeal flour, especially if finely ground, and is attributed to lipase(s) located in the outer bran layers.[15,22–24] Lipase activity is maximal at 0.8 water activity (ca. 17% moisture content),[15,25,26] but is still detectable at 0.15–0.20 water activity.[27] FFA content of wheat or flour lipids is an index of lipid deterioration,[26] but not necessarily of a loss in baking quality.[28,29]

In white flour all classes of NSL are subject to slow hydrolysis and, since there is no accumulation of monoacylglycerides, there must be some complete deacylation to FFA and water-soluble products such as glycerol, mono- and di-galactosylglycerol and glycerolphosphate esters.[30] Hydrolysis of flour lipids is caused by lipase (as in wholemeal flour) and by a polar lipid acyl hydrolase (PLAH). Lipase acts at the oil–water interface of sphero-somes and oil droplets and the principal substrate is TG, whereas PLAH acts on all classes of polar glycerolipids (but not TG), presumably in the monodisperse state.[2,26,31] The threshold for PLAH is about 0.55 water activity.

Autoxidation of lipids during prolonged storage is only a problem in flour at low-moisture contents.[32] Small selective losses of linoleate and linolenate, indicating lipoxygenase activity, have been reported, and there may be some re-esterification of free sterols with FFA,[32] but these processes are insignificant compared with lipid hydrolysis.

Lipid hydrolysis affects flour baking quality adversely in several ways. Firstly, free linoleic and linolenic acids are substrates for wheat lipoxygenase (see below), and given high levels of FFA, the enzyme consumes much of the available oxygen in dough that might be used by other oxidase enzymes.[33] Secondly, hydrolysis of PoL at a faster rate than NL adversely changes the balance of NL:PoL (see section 9.2.3) and this should decrease loaf volume. Thirdly, unsaturated FFAs weaken gluten[29,34] and seriously impair the foaming properties of the aqueous phase of dough,[35,36] giving poorer gas retention and oven spring.

9.2.2 Lipoxygenase

Lipoxygenases (LOX) require as substrates oxygen and polyunsaturated FA with the pentadiene structure in the n-6 or n-3 position (i.e. linoleate and linolenate in wheat). Two types of plant LOX are recognized. Type I LOX creates hydroperoxy radicals which are rapidly converted into more stable hydroperoxides and other products, with little formation of volatiles and negligible coupled secondary oxidations. Type II LOX is more reactive, forming volatiles by decomposition of oxidized FAs 18:2 and 18:3, and causing coupled secondary oxidations that bleach carotenoids and modify proteins.[2,37]

Wheat LOX consists of three type I isoenzymes[38-40] that use FFA as their principal substrate, and also have low activity towards MG.[41-43] It consumes most of the oxygen in unyeasted doughs and batters rapidly, but in leavened doughs it competes with yeast (and other oxidase enzymes) for available oxygen. The enzyme is concentrated in the scutellum and embryo;[15,39] activity is therefore greatest in wholemeal flour and in those millstreams that contain germ fragments[44] and high levels of FFA (see section 9.2.1). It has little effect on the rheological properties of dough or on loaf quality.

Soya has a type I LOX with an optimum pH at 8.5–9.0, and two forms of type II LOX with optima at pH 6.5[33,37] which are active in dough where they oxidize linoleate and linolenate as FFA and also when esterified in any of the classes of lipid found in flour NSL.[43] Soya flour is used as a source of type II LOX since it causes significant bleaching of flour carotenoids, and it improves the mixing tolerance and rheological properties of dough, and bread quality.[45-47] High levels of bean (*Vicia faba*) LOX can cause off flavours.[38]

9.2.3 Free and bound lipids – composition and properties

Free lipids are extracted from flour or freeze-dried dough with a non-polar solvent such as hexane, light petroleum or diethyl ether. In practice they are always part of the non-starch lipids, and the remainder, the bound lipids, can be recovered by extraction with polar solvents such as water-saturated n-butanol or chloroform–methanol–water.[48,49] The lipids inside starch granules are much more strongly bound and have very limited functions in baking (see section 9.2.7). The definition of free lipids is somewhat arbitrary, and the quantity and composition of free lipids extracted from flour depends on solvent polarity, flour moisture and particle size.[50]

Some free lipids in flour become bound when its moisture content is

Table 9.3 Composition of lipid in flour and doughs (mg/100 g dry wt)[a]

	Flour NSL		Free lipids in dough	
Lipid class	Total	Free	Anaerobic	Aerobic
TG	692	656	246	325
FFA	110	78	11	9
other NG	168	156	105	103
MGDG	80	–	–	–
DGDG	295	–	–	–
other GL	155	26	12	17
diacyl PL	222	171	8	10
lyso PL	187			
Total	1909	1067	382	464

[a] From Mann and Morrison, ref. 42.

increased to 36–45%; at this stage there is sufficient water to allow formation of a cohesive gluten structure, and dough mixing then promotes further lipid binding.[51] Under anaerobic conditions lipid binding increases markedly with the energy expended on mixing, but decreases slightly under aerobic conditions, and when air is admitted to an anaerobic dough it reverts irreversibly to the level of free lipids in an aerobic dough.[51] Soya LOX promotes release of some bound TG in dough.[45] The composition of the free lipids in flour, and from anaerobic and aerobic doughs mixed to a low work level, is given in Table 9.3.

Hard shortening fats remain largely free in dough[52,53] while surfactant dough conditioners (emulsifiers), which are similar in polarity to the principal wheat GL, become bound. Numerous variations and minor exceptions to these general statements can be found in the literature,[29] possibly reflecting responses to different dough formulae and mixing conditions. High-HLB surfactant dough conditioners, and wheat GL or DGDG, are very effective for improving the volume of bread made from protein-enriched composite flours,[29] and it is thought that they act in a similar manner to the free PoL in flour (see section 9.2.4).

Experiments with solvent-extracted flours reconstituted with free lipids or with bound lipids have shown that free lipids have a more beneficial effect on loaf volume.[54] When added to chloroform-extracted flour, small amounts of PoL are detrimental but at higher levels they are very beneficial, whereas NL are consistently detrimental.[35,36,55] In several studies significant negative correlations have been reported between the ratio of free NL/free PoL and loaf volume or (better) loaf volume corrected to a standard protein content.[56–61] However, others have failed to find any consistent correlations,[62–67] which suggests that in the latter studies other factors (e.g. gluten proteins) were a much greater source of variation between samples.

While most research has focused on the role of free lipids in breadmaking, effects have also been demonstrated on the quality of cookies, cakes and pasta.[29,35,68]

9.2.4   *Lipid binding – older theories and observations*

From various lines of experimental evidence a general model was proposed to explain the beneficial effects of free PoL and high-HLB surfactants.[29,69] It was suggested that these lipids form predominantly hydrophobic bonds with glutenins and hydrophilic bonds with gliadins to create a cross-linked and stronger gluten structure. In a similar manner GL or surfactants could integrate soya and other non-wheat protein supplements into the gluten structure.

The concept of specific lipid–protein interactions led to searches for lipid-binding proteins, and several have been isolated and characterized. Lipopurothionin is a complex of globulin protein and PoL extracted (with free lipids) from flour using light petroleum.[70,71] ligolin is a protein that binds unsaturated TG such as glyceroltrioleate but not the saturated type of TG found in hard fat,[72] S-protein is a component of gliadin-type protein aggregated by lipid[73,74] and C–M-type proteins are extracted with solvents based on chloroform and methanol.[75–79]

None of these proteins has any demonstrated functionality in bread doughs, and the validity of the claimed specific interactions between proteins and lipids has been challenged.[68,80–83] It now seems fairly certain that most, if not all, of the lipoprotein complexes are artefactual and that the nature of the proteins can vary when alternative methods are used for their isolation.[68]

9.2.5   *Lipid binding – current concepts*

In recent years completely different explanations of lipid binding and functionality have evolved based on observations of the polymorphic behaviour of polar lipids dispersed in an aqueous phase. In the presence of water, wheat NSL form various mesophasic structures.[80,84–86] Using high-resolution electron microscopy and ^{31}P-NMR to study lipids in grain, flour and dough, it has been shown that wheat NSL exhibit the characteristic polymorphic forms expected for the different water contents typical of these products.[87–89] A new consensus is now emerging on the nature of free and bound lipids.[68,81–83]

In developing grain PoL will be structural components of lipoprotein membranes in which there is either a classical PoL bilayer, or a PoL monolayer at the bounding membrane of spherosomes. As the grain matures and dries some membranes dissemble, their proteins fusing with the storage proteins (prolamins) and cytoplasmic protein matrix, while the lipids form hexagonal arrays (Figure 9.1). At this stage intact bilayer

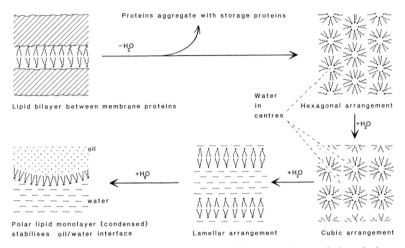

Figure 9.1 Changes in the diacyl polar lipids from the bilayer membranes of wheat during grain maturation and desiccation (top), and subsequent hydration in milled flour (right) leading to stabilization of oil droplets in water (bottom), one sequence of possible events discussed in the text. If the polar lipids are stabilizing an aqueous foam, the OIL droplet (bottom left) could be depicted as a GAS bubble, and the interfacial layer could be a condensed monolayer of polar lipids (as shown) or a mixed expanded layer of polar lipids with protein.

membrane lipids will be bound, while spherosome lipids and PoL hexagonal arrays will be mostly extractable and free. The postulated degeneration of bilayer membranes is perhaps part of the senescent processes that occur in maturing grain. Free PoL levels are controlled by two genes, *Fpl-1* and *Fpl-2* on chromosome 5D,[59] the latter being indistinguishable from the gene *Ha/ha* that regulates endosperm softness/hardness. Hence, higher levels of free PoL can be associated with grain softness which may also be a consequence of senescent changes affecting endosperm structure.

On hydration the PoL transform via the cubic phase to a lamellar phase which can then provide monodisperse PoL molecules which accumulate at the aqueous interface of oil or fat droplets to stabilize them (Figure 9.1). Lamellar phase PoL may also transform into liposomes (concentric spherical shells of lamellar-type PoL/water structures) or an inverse lamellar (L-2) phase,[80] or they could move to the aqueous interface of gas bubbles (see section 9.2.7).

Lipid binding on hydration of flour would ensue when the hydrated polar ends of PoL are external to residues of vesicles and other structures, and the FA ends are therefore inaccessible to the water-imiscible non-polar solvents used to extract free lipids. Another likely cause of binding is physical entrapment of oil or fat droplets, spherosomes and liposomes within the gluten matrix. Thus, the integrity of the matrix will depend on the work input on mixing and on redox conditions,[45,51] and the principal lipid affected by varying these factors is TG.[42] As yet, there is no convincing description of the behaviour of high-melting hard shortening fat which remains free in dough.[68] No specific protein–lipid interactions need exist (other than in remnants of the native membranes) to explain binding of NL or PoL.

9.2.6   Starch–lipid interactions

In the large A-type starch granules of wheat and barley, which comprise the bulk of the total starch, amylopectin (AP), amylose (AM) and lipids are asymmetrically distributed[90,91] so that on gelatinization the granules swell (below) and buckle into unique 'saddle' shapes.[92] This may relate to the fact that only other *Triticeae* starches can substitute satisfactorily for wheat starch in breadmaking.[29] Starch gelatinization temperature (GT) and lipid content both increase (independently) when temperatures are higher during the grain-filling period,[10] hence edaphic variation can affect GT and lipid content in the starches of various types of wheat and barley.[8,9]

Gelatinization of starch, meaning disordering of partially crystalline regions of AP, begins about 7–10°C below GT and ends about 10–15°C above GT, which is typically 56–65°C, for wheat starch in excess water.[9,93] Significant swelling of starch granules (and hence absorption of free water from dough during baking) commences at the start of gelatinization and continues to temperatures above 85°C.[92] The extent of swelling at temperatures above GT depends on AP content (which is fairly constant in wheat) and on lipids which strongly inhibit swelling, probably in the form of insoluble inclusion complexes with some of the AM.[10,92] Monoacyl lipids, such as FFA and MG from the NSL, and added AM-complexing surfactants, could have a similar effect to the starch LPL, and starch–lipid interactions may influence the loaf volume ultimately achieved during baking (see section 9.2.7). Analysis of data for starches from field-grown wheats[9] and barleys[8] showed positive correlations between solar radiation (or accumulated temperature) and LPL content which, in turn, was negatively correlated with swelling factor at 70°C.

9.2.7   Gas retention in dough

The expansion of dough during the later stages of proof, when the gluten matrix begins to show discontinuities,[94] and during baking clearly depends on its ability to retain gas under slight pressure, and this has focused attention on the foaming properties and stability of the free aqueous phase in dough (about 36–40% water, based on flour weight, is bound to protein, damaged starch and cell wall polysaccharides and so cannot participate in foaming). Foam stability of the aqueous phase obtained by centrifuging flour–water mixtures is affected by flour lipids, fractions with a high FFA content being particularly detrimental.[35] The primary substance stabilizing the dough foam has not been identified, but is probably protein, and in theory lipids could either stabilize or destabilize the foam depending on their structure.[35] Foaming capacity increases with fermentation time and LOX activity,[95] and changes in the composition of the lipids in the aqueous phase also may be relevant.[96]

It is widely supposed that wheat PoL (DGDG and MGDG in particular) stabilize gas bubbles by forming a monolayer at the gas/liquid interface.[68,80,81,84–86,94] Some experiments show that proteins do not form mixed films with PoL that form condensed monolayers, although mixed films can be formed with PoL that give expanded monolayers.[80,86,97,98] The acyl chains in wheat PoL are highly unsaturated (Table 9.1), and would be expected to give expanded monolayers, but it is not known how they interact with NL and with the proteins thought to stabilize the foam in dough. GL and surfactants with unsaturated FA are less effective than those with saturated FA for producing foams in water, and for improving gas bubble dispersion in batters.[99] Hence the question of whether or not wheat PoL stabilize dough foams effectively still remains to be answered.

Oils can impair protein foams, and wheat TG (or possibly most of the NL) might also have this effect[99] unless stabilized with a monolayer of PoL as in spherosomes (Figure 9.1). FFA are probably the most detrimental lipids;[34,35] their mode of action is not known but it could be to alter the polymorphic behaviour of the PoL[80,84] or they might aggregate the proteins that create the foam film in the first instance. The adverse effects of FFA can be offset by adding fat to dough,[34,53] which may absorb or 'sequester' FFA.

High-melting fat improves the gas retention and oven spring of bread during the earlier stages of baking,[53,100,101] and also during the expansion under vacuum of dough at proof temperature.[102] This type of fat remains essentially free in dough, but there is no completely satisfactory explanation of the mechanism(s) that may operate.[53,68,100–102]

There are two less obvious ways in which starch and lipids could affect dough rheology and expansion during baking. When starch granules are mechanically damaged they absorb water and swell much more at ambient temperature ('cold gelatinization') than undamaged granules do, and this well-known property is used by millers to regulate flour water-absorption. However, at any given level of damage, starch swelling is negatively correlated with LPL content (R. F. Tester and W. R. Morrison, unpublished observations) and appreciable differences in swelling and water uptake could occur as a result of variation in starch (see section 9.2.6).

The second situation arises when starch granules gelatinize and swell during baking so that eventually they remove all of the foam-forming free aqueous phase from the dough. Since this is one of the events that 'sets' the structure of the loaf it does not collapse. Clearly, if gelatinization and swelling occur at temperatures several degrees higher than with a low-GT starch, oven spring can continue longer during baking. Test loaves baked using flours reconstituted with starch from various types of wheat, showed that loaf specific volume was significantly correlated with starch GT, but not with starch lipid alone,[93] although lipid was almost certainly a factor contributing to variation in starch swelling and loaf volume.

9.3 Summary

The wheat grain contains 2–3% lipids which have structural and storage functions characteristic of the tissues in which they occur. The lipids are a useful source of *n*-6 and *n*-3 types of polyunsaturated fatty acids, and wheat germ is exceptionally rich in vitamin E tocopherols. Milling redistributes aleurone and germ lipids among millstreams, where they become more susceptible to lipolysis when the flour is stored. Some oxidation of lipids by wheat lipoxygenase occurs in dough, but more important oxidations resulting in dough improvement require soya lipoxygenase, and both contribute to the oxygen requirement of the dough. Polar and non-polar lipids affect gas retention and expansion of dough in the oven. Older theories to explain the observed effects of free and bound lipids invoked specific lipid–protein interactions which no longer seem likely. Current concepts explain the polymorphic transitions of wheat lipids in the same terms as for other biological systems, and it is thought that the lipids affect loaf volume by altering the foam stability (and hence gas retention) of the free aqueous phase in dough.

References

1. Morrison, W. R. (1978) Cereal lipids. *Adv. Cereal Sci. Technol.*, **2**, 221–348.
2. Morrison, W. R. (1988) Lipids. In *Wheat: Chemistry and Technology*, Vol. 1, (ed. Y. Pomeranz) American Association of Cereal Chemists, St. Paul, MN, pp. 373–439.
3. Morrison, W. R. (1988) Lipids in cereal starches: a review. *J. Cereal Sci.*, **8**, 1–15.
4. Barnes, P. J. (1983) Non-saponifiable lipids in cereals. In *Lipids in Cereal Technology*, (ed. P. J. Barnes) Academic Press, London, pp. 33–55.
5. Hargin, K. D., Morrison, W. R. and Fulcher, R. G. (1980). Triglyceride deposits in the starchy endosperm of wheat. *Cereal Chem.*, **57**, 320–325.
6. Morrison, W. R. (1981) Starch lipids: a reappraisal. *Starch/Stärke*, **33**, 408–410.
7. Hargin, K. D. and Morrison, W. R. (1980) The distribution of acyl lipids in the germ, aleurone, starch and non-starch endosperm of four wheat varieties. *J. Sci. Food Agric.*, **32**, 877–888.
8. Morrison, W. R., Scott, D. C. and Karkalas, J. (1986) Variation in the composition and physical properties of barley starches. *Starch/Stärke*, **38**, 374–379.
9. Morrison, W. R. (1989) Uniqueness of wheat starch, in *Wheat is Unique*, (ed. Y. Pomeranz) American Association of Cereal Chemists, St. Paul, Minnesota, pp. 193–214.
10. Tester, R. F., South, J. B., Morrison, W. R. and Ellis, R. P. (1991) The effects of ambient temperature during the grain filling period on the composition and properties of starches from four barley genotypes. *J. Cereal Sci.*, **13**, 113–127.
11. Stevens, D. J. (1959) The contribution of the germ to the oil content of white flour. *Cereal Chem.*, **36**, 452–461.
12. Morrison, W. R. and Hargin, K. D. (1981) Distribution of soft wheat kernel lipids into flour milling fractions. *J. Sci. Food Agric.*, **32**, 579–587.

13. Morrison, W. R., Coventry, A. M. and Barnes, P. J. (1982) The distribution of acyl lipids and tocopherols in flour millstreams. *J. Sci. Food Agric.*, **33**, 925–933.
14. Morrison, W. R. and Barnes, P. J. (1983) Distribution of wheat acyl lipids and tocols in millstreams. In *Lipids in Cereal Technology*, (ed. P. J. Barnes) Academic Press, London, pp. 149–164.
15. Galliard, T. (1986) Wholemeal flour and baked products: chemical aspects and functional properties. In *Chemistry and Physics of Baking*, (eds J. M. V. Blanshard, P. J. Frazier and T. Galliard) Royal Society of Chemistry, London, pp. 199–215.
16. Makarov, V. V., Prokhorova, A. P., Razorenova, E. E. and Chutkova, E. I. (1974) Experiments in long-term storage of wheat in industrial conditions. *Tr. Vses. Nauchno Issled. Inst. Zerna Prod. Ego Pererab.*, **80**, 72–78.
17. Pixton, S. W., Warburton, S. and Hill, S. T. (1974) Long-term storage of wheat. III. Some changes in the quality of wheat observed during 16 years storage. *J. Stored Prod. Res.*, **11**, 177–185.
18. Daftary, R. D. and Pomeranz, Y. (1965) Changes in lipid composition in wheat during storage deterioration. *J. Agric. Food Chem.*, **13**, 442–446.
19. Daftary, R. D., Pomeranz, Y. and Sauer, D. B. (1970) Changes in wheat flour damaged by mold during storage. Effect on lipid, lipoprotein and protein. *J. Agric. Food Chem.*, **18**, 613–616.
20. Quaglia, G., Cavaioli, R., Catani, P., Shejbal, J. and Lombardi, M. (1980) Preservation of chemical parameters in cereal grains stored in nitrogen, in *Controlled Atmosphere Storage of Cereal Grains*, (ed. J. Shejbal) Elsevier, Amsterdam, pp. 319–333.
21. El Baya, A. W., Fretzdorff, B. and Muenzing, K. (1986) The behaviour of lipids during storage of wheat under controlled atmosphere. *Getreide Mehl Brot*, **40**, 71–78.
22. Barnes, P. J. and Lowy, G. D. A. (1986) The effect on quality of interaction between milling fractions during the storage of wheat flour. *J. Cereal Sci.*, **4**, 225–232.
23. Galliard, T. (1986) Oxygen consumption of aqueous suspensions of whole wheatmeal, bran and germ: involvement of lipase and lipoxygenase. *J. Cereal Sci.*, **4**, 33–50.
24. Galliard, T. (1986) Hydrolytic and oxidative degradation of lipids during storage of wholemeal flour: Effects of bran and germ components. *J. Cereal Sci.*, **4**, 179–182.
25. Drapron, R. (1972) Enzymic reactions in low-moisture systems. *Ann. Technol. Agric.*, **21**, 487–499.
26. Galliard, T. (1983) Enzymic degradation of cereal lipids. In *Lipids in Cereal Technology*, (ed. P. J. Barnes) Academic Press, London, pp. 111–148.
27. Acker, L. and Beutler, H.-O. (1965) Enzymic fat splitting in low moisture foods. *Fette Seifen Anstrichm.*, **67**, 430–433.
28. Bell, B. M., Chamberlain, N., Collins, T. H., Daniels, D. G. H. and Fisher, N. (1979) The composition, rheological properties and breadmaking behaviour of stored flours. *J. Sci. Food Agric.*, **30**, 1111–1122.
29. Pomeranz, Y. (1988) Composition and functionality of wheat flour components, in *Wheat: Chemistry and Technology*, Vol. 2, (ed. Y. Pomeranz) American Association of Cereal Chemists, St. Paul, Minnesota, pp. 219–370.
30. Clayton, T. A. and Morrison, W. R. (1972) Changes in wheat flour lipids during the storage of wheat flour. *J. Sci. Food Agric.*, **23**, 721–736.
31. Galliard, T. (1983) Assays for lipid-degrading enzymes. In *Lipids in Cereal Technology*, (ed. P. J. Barnes) Academic Press, London, pp. 403–408.
32. Shearer, G. and Warwick, M. J. (1983) The effect of storage on lipids and breadmaking properties of wheat flour, in *Lipids in Cereal Technology*, (ed. P. J. Barnes) Academic Press, London, pp. 253–268.
33. Grosch, W. (1986) Redox systems in dough. In *Chemistry and Physics of Baking*, (eds J. M. V. Blanshard, P. J. Frazier and T. Galliard) Royal Society of Chemistry, London, pp. 155–169.
34. De Stefanis, V. A. and Ponte, J. G. (1976) Studies on the breadmaking properties of wheat flour nonpolar lipids. *Cereal Chem.*, **53**, 636–642.
35. MacRitchie, F. (1983) Role of lipids in baking. In *Lipids in Cereal Technology*, (ed. P. J. Barnes) Academic Press, London, pp. 165–188.
36. MacRitchie, F. and Gras, P. W. (1973) The role of flour lipids in baking. *Cereal Chem.*, **50**, 292–302.
37. Grosch, W., Laskawy, G. and Weber, F. (1976) Formation of volatile carbonyl

compounds and co-oxidation of β-carotene by lipoxygenase from wheat, potato, flax and beans. *J. Agric. Food Chem.*, **24**, 456–459.

38. Nicolas, J. and Drapron, R. (1983) Lipoxygenase and some related enzymes in breadmaking. In *Lipids in Cereal Technology*, (ed. P. J. Barnes) Academic Press, London, pp. 213–236.

39. Galliard, T. (1983) Enzymic oxidation of wheat flour lipids. In *Developments in Food Science, Vol. 5A. Progress in Cereal Chemistry and Technology*, (ed. J. Holàs) Elsevier, Amsterdam, pp. 419–424.

40. Shiiba, K., Negishi, Y., Okada, K. and Nagao, S. (1991) Purification and characterization of lipoxygenase isoenzymes from wheat germ. *Cereal Chem.*, **68**, 115–122.

41. Graveland, A. (1970) Enzymatic oxidations of linoleic acid and glycerol-1-monolinoleate in doughs and flour–water suspensions. *J. Am. Oil Chem. Soc.*, **47**, 352–361.

42. Mann, D. L. and Morrison, W. R. (1974) Changes in wheat flour lipids during mixing and resting of flour-water doughs. *J. Sci. Food Agric.*, **25**, 1109–1119.

43. Morrison, W. R. and Panpaprai, R. (1975) Oxidation of free and esterified linoleic and linolenic acids in bread doughs by wheat and soya lipoxygenases. *J. Sci. Food Agric.*, **26**, 1225–1236.

44. Von Ceumern, S. and Hartfiel, W. (1984) Activity of lipoxygenase in cereals and possibilities of enzyme inhibition. *Fette Seifen Anstrichm.*, **86**, 204–208.

45. Frazier, P. J. (1979) Lipoxygenase action and lipid binding in dough. *Bakers Digest*, **53**(6), 8–10, 12, 13, 16, 18, 20, 29.

46. Hoseney, R. C., Rao, H., Faubion, J. and Sidhu, J. S. (1980) Mixograph studies. IV. The mechanism by which lipoxygenase increases mixing tolerance. *Cereal Chem.*, **57**, 163–166.

47. Kieffer, R. and Grosch, W. (1980) Improvement of the baking properties of wheat flour by type II lipoxygenase from soya beans. *Z. Lebensm. Unters. Forsch.*, **170**, 258–261.

48. Morrison, W. R., Mann, D. L., Wong, S. and Coventry, A. M. (1975) Selective extraction and quantitative analysis of non-starch and starch lipids from wheat flour. *J. Sci. Food Agric.*, **26**, 507–521.

49. Morrison, W. R., Tan, S. L. and Hargin, K. D. (1980) Methods for the quantitative analysis of lipids in cereal grains and similar tissues. *J. Sci. Food Agric.*, **31**, 329–340.

50. Chung, O. K., Pomeranz, Y., Jacobs, R. M. and Howard, B. G. (1980) Lipid extraction conditions to differentiate among hard red winter wheats that vary in breadmaking. *J. Food Sci.*, **45**, 1168–1174.

51. Daniels, N. W. R. (1975) Some effects of water in wheat flour doughs. In *Water Relations of Foods*, (ed. R. B. Duckworth) Academic Press, London, pp. 573–586.

52. Bechtel, D. B., Pomeranz, Y. and de Franciso, A. (1978) Breadmaking study by light and electron microscopy. *Cereal Chem.*, **55**, 392–401.

53. Bell, B. M., Daniels, D. G. H. and Fisher, N. (1979) The effects of pure saturated and unsaturated fatty acids on breadmaking and on lipid binding, using Chorleywood Bread Process doughs containing a model fat. *J. Sci. Food Agric.*, **30**, 1123–1130.

54. Hoseney, R. C. and Finney, K. F. (1971) Functional (breadmaking) and biochemical properties of wheat flour components. XI. A review. *Bakers Digest*, **45**(4), 30–36, 39–40, 64–67.

55. McCormack, G., Panozzo, J. and MacRitchie, F. (1991) Contributions to breadmaking of inherent variations in lipid content and composition of wheat cultivars. II. Fractionation and reconstitution studies. *J. Cereal Sci.*, **13**, 263–274.

56. Chung, O. K., Pomeranz, Y. and Finney, K. F. (1982) Relation of polar lipid content to mixing requirement and loaf volume potential of hard red winter wheat flour. *Cereal Chem.*, **59**, 14–20.

57. Zawistowska, U., Bèkès, F. and Bushuk, W. (1984) Intercultivar variations in lipid content, composition and distribution and their relation to baking quality. *Cereal Chem.*, **61**, 527–531.

58. Bèkès, F., Zawistowska, U., Zillman, R. R. and Bushuk, W. (1986) Relationship between lipid content and composition and loaf volume of twenty-six common spring wheats. *Cereal Chem.*, **63**, 327–331.

59. Morrison, W. R., Law, C. N., Wylie, L. J., Coventry, A. M. and Seekings, J. (1989) The effect of group 5 chromosomes on the free polar lipids and breadmaking quality of wheat. *J. Cereal Sci.*, **9**, 41–51.

60. Karpati, E. M., Bèkès, F., Lasztity, R., Örsi, F., Smied, I. and Mosonyi, A. (1990) Investigation of the relationship between wheat lipids and baking properties. *Acta Aliment.*, **19**, 237–260.
61. Matsoukas, N. P. and Morrison, W. R. (1991) Breadmaking quality of ten Greek breadwheats. II. Relationships of protein, lipid and starch components to baking quality. *J. Sci. Food Agric.*, **55**, 87–101.
62. Berger, M. (1983) Soft wheat lipids 2. Composition of free and bound lipids in flour from eight varieties of French soft winter wheat. *Sci. Aliment.*, **3**, 181–217.
63. Larsen, N. G., Baruch, D. W. and Humphrey-Taylor, V. J. (1986) Lipid quality factors in breeding New Zealand wheats. *Agron. Soc. NZ*, spec. publ. **50**, 278.
64. Larsen, N. G., Humphrey-Taylor, V. J. and Baruch, D. W. (1989) Glycolipid content as a breadmaking quality determinant in flours from New Zealand wheat blends. *J. Cereal Sci.*, **9**, 149–157.
65. Bell, B. M., Daniels, D. G. H., Fearn, T. and Stewart, B. A. (1987) Lipid composition, baking quality and other characteristics of wheat varieties grown in the UK. *J. Cereal Sci.*, **5**, 277–286.
66. Marston, P. and MacRitchie, F. (1985) Lipids: Effects on the breadmaking quality of Australian flour. *Food Technol. Aust.*, **37**, 362–365.
67. McCormack, G., Panozzo, J., Bèkès, F. and MacRitchie, F. (1991) Contributions to breadmaking of inherent variations in lipid content and composition of wheat cultivars. I. Results of survey. *J. Cereal Sci.*, **13**, 255–261.
68. Carr, N. O. (1991) Lipid binding and lipid-protein interaction in wheat flour dough. *PhD thesis*, University of Reading, Reading.
69. Chung, O. K. (1986) Lipid–protein interactions in wheat flour, dough, gluten and protein fractions. *Cereal Foods World*, **31**, 242–244, 246–247, 249–252, 254–256.
70. Lasztity, R. (1984) Wheat proteins, in *The Chemistry of Cereal Proteins*, (ed. R. Lasztity) CRC Press, Boco Raton, pp. 79–86.
71. Garcia-Olmedo, F., Rodriguez-Palenzuela, P., Hernandez-Lucas, C., Ponz, F., Marana, C., Carmona, M. J., Lopez-Fando, J., Fernandez, J. A. and Carbonero, P. (1989) The thioxins: A protein family that includes purothionins, viscotoxins and crambins, in *Oxford Surveys of Plant Molecular Biology*, Vol. 6, (ed. B. J. Miflin) Oxford University Press.
72. Frazier, P. J., Daniels, N. W. R. and Russell-Eggitt, P. W. (1981) Lipid–protein interactions during dough development. *J. Sci. Food Agric.*, **32**, 877–897.
73. Zawistowska, U., Bietz, J. A. and Bushuk, W. (1986) Characterization of low-molecular weight protein with high affinity for flour lipid from two classes of wheat. *Cereal Chem.*, **63**, 414–419.
74. Zawistowska, U. and Bushuk, W. (1986) Electrophoretic characterization of low-molecular weight wheat protein of variable solubility. *J. Sci. Food Agric.*, **37**, 409–417.
75. Meredith, P. (1965) On the solubility of gliadin-like proteins. *Cereal Chem.*, **42**, 54–63, 64–148, 149–160.
76. Belova, T. E., Dubtsova, G. N., Kolesnik, G. B., Doronina, O. D. and Nechaev, A. P. (1981) Lipoprotein complex of wheat grain. *Prikl. Biokhim. Mikrobiol.*, **17**, 734–738.
77. Garcia-Olmedo, F., Carbonero, P. and Jones, B. L. (1982) Chromosomal locations of genes that control wheat endosperm proteins. *Adv. Cereal Sci. Technol.*, **5**, 1–47.
78. Kobrehel, K. and Sauvaire, Y. (1990) Particular lipid composition in isolated proteins of durum wheat. *J. Agric. Food Chem.*, **38**, 1164–1171.
79. Trusova, V. M., Zhumanova, U. T. Mochanov, M. I. and Nechaev, A. P. (1990) Selective binding of glycolipids to lipoproteins of the wheat gluten complex. *Prikl. Biokhim. Mikrobiol.*, **26**, 789–793.
80. Larsson, K. (1986) Functionality of wheat lipids in relation to gluten gel formation, in *Chemistry and Physics of baking*, (eds J. M. V. Blanshard, P. J. Frazier and T. Galliard) Royal Society of Chemistry, London, pp. 61–74.
81. Marion, D., Le Roux, C., Tellier, C., Akoka, S., Gallant, D., Gueguen, J., Popineau, Y. and Compoint, J. P. (1989) Lipid–protein interactions in wheat gluten: A renewal. *Abh. Akad. Wiss. DDR, Abt. Math., Naturwiss., Tech.* 1N, *Interactions in Protein Systems*, pp. 147–152 (see also *Chem. Abstr.* 113(7) 057566).
82. Morrison, W. R. (1989) Wheat lipids are unique, in *Wheat is Unique*, (ed. Y. Pomeranz) American Association of Ceral Chemistry, St. Paul, Minnesota, pp. 319–339.

83. Morrison, W. R. (1989) Recent progress on the chemistry and functionality of flour lipids, in *Wheat End-Use Properties*, (ed. H. Salovaara) Proceedings, ICC '89 Symposium, Lahti, Finland, University of Helsinki, pp. 131–149.

84. Carlson, T., Larsson, K. and Miezis, Y. (1978). Phase equilibria and structure in the aqueous system of wheat lipids. *Cereal Chem.*, **55**, 168–179.

85. Carlson, T., Larsson, K., Miezis, Y. and Poovaradom, S. (1979) Phase equilibria in the aqueous system of wheat gluten lipids and in the aqueous salt system of wheat lipids. *Cereal Chem.*, **56**, 417–419.

86. Larsson, K. (1983) Physical state of lipids and their technical effects in baking, in *Lipids in Cereal Technology*, (ed. P. J. Barnes) Academic Press, London, pp. 237–252.

87. Al Saleh, A., Marion, D. and Gallant, D. (1986) Microstructure of mealy and vitreous wheat endosperms (*Triticum durum* L.) with special emphasis on location and polymorphic behaviour of lipids. *Food Microstruct.*, **5**, 131–140.

88. Akoka, S., Tellier, C., Le Roux, C. and Marion, D. (1988) A phosphorus magnetic resonance spectroscopy and a differential scanning calorimetry study of the physical properties of N-acylphosphatidylethanolamines in aqueous dispersions. *Chem. Phys. Lipids*, **46**, 43–50.

89. Marion, D., Le Roux, C., Akoka, S., Tellier, C. and Gallant, D. (1987) Lipid–protein interaction in wheat gluten: A phosphorus nuclear magnetic resonance spectroscopy and freeze-fracture electron microscopy study. *J. Cereal Sci.*, **5**, 101–115.

90. Morrison, W. R. and Gadan, H. (1987) The amylose and lipid contents of starch granules in developing wheat endosperm. *J. Cereal Sci.*, **5**, 263–275.

91. McDonald, A. M. L., Stark, J. R., Morrison, W. R. and Ellis, R. P. (1991) The composition of starch granules from developing barley genotypes. *J. Cereal Sci.*, **13**, 93–112.

92. Tester, R. F. and Morrison, W. R. (1990) Swelling and gelatinization of cereal starches. I. Effects of amylopectin, amylose and lipids. *Cereal Chem.*, **67**, 551–557.

93. Soulaka, A. B. and Morrison, W. R. (1985) The bread baking quality of six wheat starches differing in composition and physical properties. *J. Sci. Food Agric.*, **36**, 719–727.

94. Gan, Z., Angold, R. E., Williams, M. R., Ellis, P. R., Vaughan, J. G. and Galliard, T. (1990) The microstructure and gas retention of bread dough. *J. Cereal Sci.*, **12**, 15–24.

95. Shiiba, K., Negishi, Y., Okada, K. and Nagao, S. (1990) Chemical changes during sponge-dough fermentation. *Cereal Chem.*, **67**, 350–355.

96. Tortosa, E., Ortola, C. and Barber, S. (1986) Chemical changes during bread dough fermentation. VI. Lipids in the liquid phase of bread dough. *Rev. Agroquim. Tecnol. Aliment.*, **26**, 269–275.

97. Cornell, D. G. (1982) Lipid–protein interactions in monolayers: egg yolk phosphatidic acid and β-lactoglobulin. *J. Colloid Interface Sci.*, **88**, 536–545.

98. Cornell, D. G. and Carroll, R. J. (1985) Miscibility of lipid–protein monolayers. *J. Colloid Interface Sci.*, **108**, 226–233.

99. MacRitchie, F. (1986) Physiochemical processes in mixing. In *Chemistry and Physics of Baking*, (eds J. M. V. Blanshard, P. J. Frazier and T. Galliard) Royal Society of Chemistry, London, pp. 132–146.

100. Bell, B. M., Daniels, D. G. H. and Fisher, N. (1977) Physical aspects of the improvement of dough by fat. *Food Chem.*, **2**, 57–70.

101. Hoseney, R. C. (1986) Component interaction during heating and storage of baked products, in *Chemistry and Physics of Baking*, (eds J. M. V. Blanshard, P. J. Frazier and T. Galliard) Royal Society of Chemistry, London, pp. 216–226.

102. Bell, B. M., Daniels, D. G. H. and Fisher, N. (1981) Vacuum expansion of mechanically developed dough at proof temperature: effect of temperature. *Cereal Chem.*, **58**, 182–186.

10 Enzymes of sprouted wheat and their possible technological significance

J. E. KRUGER

10.1 Introduction

One of the most detrimental effects on the quality of a cereal such as wheat resulting in large economic losses each year is the occurrence of preharvest sprouting.[1] Preharvest sprouting occurs when a cereal is subjected to wet weather conditions just prior to harvest ripeness. The cereal may have a certain tolerance to this wet weather called preharvest sprout resistance which is dependent on the cultivar. White wheats, in general, are more susceptible to sprouting than red wheats. Sprouting is a germinative process to start a new plant on its way and results in major physiological changes occurring in the grain. Two of these changes are technologically very important from the food processing standpoint. One is the formation of germinative enzymes, either through *de novo* synthesis or reactivation of pre-existing latent ones. The second change is the degradation of the storage reserves of the plant *in situ*.

Germinative enzymes cause excessive degradation of the biochemical components necessary to produce a satisfactory end-product such as bread, and manifest their effects directly during processing. The influence of a particular enzyme will be dependent on the conditions that are suitable for enzyme activity to occur. The enzyme α-amylase is considered one of the most important in this regard for bread production.

Degradation of the storage reserves *in situ*, on the other hand, means that biochemically important components have been damaged before processing even begins. For example, if the gluten proteins have been damaged, it is likely that quality deterioration of bread will occur.

Although both situations can occur, direct damage by enzymes during processing is considered more important and will be the focus of this chapter. Only wheat end-products will be considered, but sprout damage can also have disastrous effects for cereals such as barley, rye and triticale.

The starting material for scientific studies on the effects of sprout damage is very complex and is probably the reason for the disparity in results found between laboratories. Thus, many studies have used laboratory germinated material which may be quite different in composition from actual field-sprouted samples. Other scientific studies have characterized their starting

Table 10.1 α-amylase levels (mg maltose/min g^{-1}) in ungerminated, degermed and various degrees of visual sprouting for different grades of Canada western red spring (CWRS) wheat (reprinted with permission from Kruger, ref. 2)

		Visual sprouting			
Ungerminated	Degermed	Degree 1	Degree 2	Degree 3	Degree 4
1 CWRS					
6.8	2.7	7.6	1.2	22	49
4.6	0	1.7	14	14	16
1.5		4.3	3.5	8.2	2508
3.6			5.4	2192	104
2 CWRS					
6.3	1.2	22	15	782	4025
30	0	17	3.0	31	1343
0	15342	2.5	4.0	69	21
0	395	1.2	8.6	12	805
3 CWRS					
5.6	4720	106	490	6505	273
14	1311	17	107	68	4387
10	19422	107	522	2313	21163
11	584	20	75	146	2116

material in terms of the percentage of sprouted kernels. However, sprouted kernels can vary enormously in the amount of an enzyme such as α-amylase that they contain, varying by as much as several thousand fold[2] (Table 10.1) The more severely sprouted a wheat kernel becomes, the more enzyme it is likely to contain. Quite often the degree of sprouting is equated to the amount of the enzyme α-amylase. The Falling Number test,[3–5] an autolytic viscometric test employing the natural starch present in ground grain as substrate, is most often used (Figure 10.1). Falling Number (*FN*) values above 300 s are considered free of sprout damage (sound), while below 160 s, wheats are generally considered unsuitable for breadmaking. Although very useful in commerce, samples with similar falling number values can be quite different in composition. The two extremes can either be a large number of very lightly sprouted kernels containing low levels of enzyme or, alternately, a small amount of highly sprouted kernels in an otherwise sound population. Effects on processing and, consequently, scientific findings could be quite different in these two situations. In the first scenario, it is likely that *in situ* damage to the storage reserves of the plant could be as important to quality deterioration as excess enzyme. In the latter case, a few severely sprouted kernels are being dealt with, but with most kernels sound and the endosperm reserves untouched. Since enzymes are synthesized during germination in the outer layer of the wheat kernel (aleurone layer) and progressively secreted into the endosperm as sprouting increases, more enzyme will end up in the flour milled from heavily sprouted kernels than from lightly sprouted ones.[6] The message from this discussion is

Figure 10.1 Falling number apparatus for measurement of sprout damage.

therefore clear. Sprout-damaged field material for research studies should be characterized as fully as possible. One should attempt to determine not only the α-amylase level of the sample and the percentage of sprouted kernels, but attempt to assess the severity of sprouted kernels that are present. Something that is useful in this regard is an instrument called the Carlsberg seed analyzer.[7] Another useful way to characterize the material is to measure quantitatively other germinative enzymes present such as proteases and polyphenol oxidase. These enzymes are synthesized at different rates to α-amylase during germination and can be of as great or greater importance in affecting the quality of certain end-products. An additional way of characterizing sprouted wheat is to investigate whether the endosperm storage reserves have indeed been damaged. This can be done using an unsprouted wheat sample for comparison and a technique such as high performance liquid chromatography (HPLC) to examine damage to the gluten proteins present in the wheat sample suspected of having sprout damage.[8–10]

There have been three approaches to solving the problem of preharvest sprouting in wheat. Plant breeders are continually trying to develop wheats which have greater resistance to sprouting under adverse wet weather conditions. Plant physiologists are examining various aspects of the physiology of the kernel under germinative conditions, hoping to find physiological means (i.e. natural inhibitors, etc.) which could prevent germinative enzymes from being produced in large amounts. Finally, cereal and food scientists are trying to establish levels of enzymes that can be tolerated for particular end-products and ameliorative ways to prevent their action during processing. It is this last approach that will be considered here with illustrations from various end-products produced from wheat flour and semolina.

10.2   Western-style breads

Preharvest sprouting damage is particularly damaging to bread products and a great deal of research effort has been devoted to examining and ameliorating the effects. The enzyme considered to have the greatest effect is α-amylase. A small amount of the enzyme is necessary in order to degrade damaged starch and produce an adequate supply of fermentable sugars for gas production, but bakers would prefer to have negligible amounts of the enzyme and add what they feel is necessary. If in excess, α-amylase breaks down the starch to complex dextrins and simple sugars, resulting in a dough that is difficult to handle, ultimately producing bread with a wet sticky crumb. The effects are most noticeable in long fermentation processes since the enzyme has more time to act. Certain types of bread are also more sprout-damage prone, such as the Pullman type of square sandwich loaf, common in Japan, where wall cave-in occurs if elevated levels of α-amylase are present.

In laboratory test baking, levels of the enzyme required to show discernible signs of quality deterioration may be greater than in commercial automated processes where the problem may progressively show up at the bread slicing stage. Difficulty in slicing, due to the buildup of dextrins on the slicing blade, will be accentuated as more loaves pass through the process. Another sign of excess sprout damage is the formation of darkened crusts due to Maillard-type reactions because of excess sugars and dextrins.

Because of its importance, the enzyme α-amylase has been well character-ized.[11–13] There are a number of isozymes of the enzyme which fall into two groups on the basis of their isoelectric points[14] (Figure 10.2). One group, called the low pI or 'green' isozymes because of their presence in immature wheat, are fairly heat labile.[15] The high pI-type or 'germinated' isozymes, on the other hand, are heat stable and consequently more active during the breadmaking process. Natural inhibitors of the enzyme have been isolated from wheat and barley and show some promise of being able to inhibit the enzyme in a bread dough situation.[16–18] At present, however, less satisfying solutions must be employed by the baker, such as decreasing the water absorption of the dough, omitting α-amylase supplementation, adding chemicals such as sodium pyrophosphate or sodium stearyl lactylate, adding extra fat and making sure that the slicer blades are well lubricated. Of course, when water absorption is lowered, the baker loses money, as he is unable to put more water into his product. That fat is effective in preventing effects due to α-amylase was demonstrated in this laboratory with the

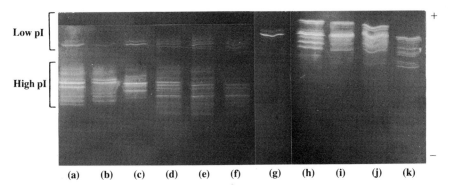

Figure 10.2 α-amylase isozymes in different cereals (reprinted with permission from MacGregor *et al.*, ref. 14). (a) 2-Rowed barley. (b) 6-Rowed barley. (c) Rye. (d) Triticale. (e) Amber Durum wheat. (f) Red Spring wheat. (g) Oats. (h) Corn. (i) Millet. (j) Sorghum. (k) Rice.

production of 'alinado', a Colombian type of French bread. The fat content of the product is around 20% and wheats with high α-amylase (*FN* values of around 100 s) have been processed with no problems (J. Dexter, personal communication).

α-Amylase should not be considered the exclusive culprit when it comes to quality deterioration of bread. Proteolytic attack, either *in situ* or during processing may also occur. Thus, with increased germination, changes in rheological properties such as farinograph absorption, dough development times and mixing tolerance indexes occur.[19–24] These are related to changes in solubility characteristics of the gluten proteins. Studies in this laboratory indicate that storage proteins become increasingly soluble in 50% 1-propanol[25,26] as germination proceeds (Figure 10.3). The solubilized proteins are initially released as aggregates rather than as small peptides and amino acids, suggesting the presence of specific endoproteolytic action.

10.3   Hearth and flat breads

An evaluation of nine hearth breads indicated that they were more tolerant to levels of sprout damage than conventional bread.[27] The products tested were Egyptian 'balady' bread, Moroccan sourdough bread, Moroccan straight dough wholewheat flour bread, Pakistani chapatti, Iranian barbari bread and four types of Indian bread. With the exception of the Iranian barbari bread and the Egyptian balady bread, processing problems could be alleviated by decreasing water absorption or adding dusting powder. Presumably the high hearth temperatures used in preparing these products minimizes the time of reaction that enzymes such as α-amylase can have before they are inactivated.

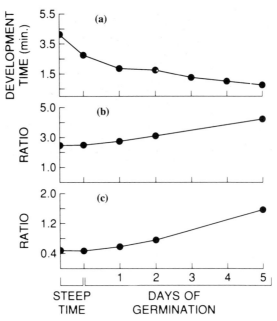

Figure 10.3 Changes during germination of HY320 wheat in (a) farinograph dough development time; (b) ratio of protein in propanol to propanol–DTT solubles as analyzed by RP-HPLC; (c) ratio of high molecular weight glutenin subunits in propanol solubles to those present in propanol–DTT soluble as analyzed by RP-HPLC (reprinted with permission from Kruger and Marchylo, ref. 25).

Flat breads are commonly produced using a higher flour extraction rate. Such products, therefore, have a greater amount of bran and as a consequence a greater amount of germinative enzymes. Studies have shown that sprout-damaged wheat flour may give rise to pocket breads having blisters or cracks on the surface, poorer crumb color and a texture that is leathery in nature and difficult to fold or roll.[28]

10.4   Chinese steamed bread

This product is widely consumed in the Orient. Formulation and fermentation conditions are somewhat like conventional bread, but with a lower water absorption (i.e. around 45%) and with the bun-shaped product being steamed for around 12–15 min rather than baked in an oven. A desirable final product should have a white shiny symmetrical crust, free of blemishes, and the crumb should be resilient to the touch. Preliminary studies that we have carried out indicate that increased levels of malt flour produced a product which became progressively flatter (Figure 10.4). The crumb also became coarser and buns were very sticky, particularly closer to the crust.

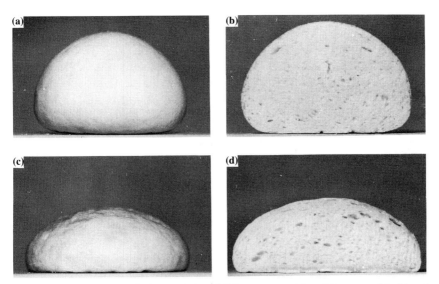

Figure 10.4 Crust and crumb characteristics of Chinese steamed bread from sound (a, b) and malted (c, d) flour.

When squeezed, the bun did not spring back. The crust and crumb color become increasingly more gray.

10.5 Spaghetti

Spaghetti is processed by combining water and durum semolina, extruding the resulting dough and drying the product by a temperature controlled process. The product should be more tolerant of sprout damage than bread production since the liquid to solid ratio is less and semolina particles are much larger than flour. As a consequence, there is considerably less mobility of enzyme and potential interaction with substrate. Upon cooking, some reaction can occur, but for a very short time before penetration of boiling water results in inactivation of enzyme. Kruger and Matsuo[29] showed that use of laboratory germinated durum wheat semolina resulted in spaghetti which had greater cooking losses compared to using ungerminated semolina. A number of other studies using field sprouted semolina have indeed shown that processing and cooking properties are not dramatically affected unless sprouting is very severe.[30–33] In a recent study[33] in which both low- and high-temperature drying were employed, only the very highly field-sprouted samples experienced any signs of quality deterioration manifested by slightly increased cooking losses and checking. Caution should be exercised in the interpretation of laboratory studies as processing problems have been experienced on a commercial scale when sprout-damaged wheat was used.

10.6 Noodles

Contrasted with spaghetti, which involves the use of a hard wheat and extrusion for processing, noodles are made with a softer wheat and by sheeting. There are a multitude of noodle types and methods of processing. In general, the formulation is quite simple, consisting of water (30–35%), flour and either salt (white salted noodles) or Kansui, an alkaline mixture of sodium and potassium carbonates (Chinese noodles). The noodles may be sold raw, dried, steamed and dried or steamed and fried. In general, very few studies have been carried out on the effects of sprout damage on noodles, but effects can be quite serious.

Japanese dry noodles are apparently very sensitive to low levels of sprout damage where the major quality defects observed are the formation of wet sticky doughs and noodles which tend to droop when drying.[34,35] Proteases, α-amylase and a heat-stable water-soluble fraction appear to be implicated, but the relative role and importance of these components have not been established.

Cantonese noodles, made with Kansui, are a pronounced yellow color caused by flavonoid pigments under alkaline conditions. Both color and texture are important with a firm but elastic product being preferred. In one study, sprout damage was shown to produce an inferior product which was darker in color and when cooked produced a soft and mushy noodle.[28] Changes to the texture are probably caused by *in situ* damage to reserves. Direct effects due to α-amylase are unlikely since the pH of the noodle dough is around 9–10, well away from the pH optimum of 5.5 for this enzyme. Very little is known, however, of the relative effects of other enzymes such as proteases, etc. Raw Cantonese noodles are more likely than dried ones to have accentuated effects due to sprout damage since the raw noodle may be stored for 24 h or longer before it is cooked. Studies that we have carried out indicate that the enzyme polyphenol oxidase (PPO) is likely to be involved in the deterioration of color.[36] Thus, flour samples with high-PPO levels produce raw Cantonese noodles which have an appreciably faster decrease in brightness coincident with an increased brown color. The same phenomena occurs when increased levels of a sprout-damaged flour are added to a sound flour. Since the PPO enzyme is located in the branny layer of the wheat kernel, the problem can be decreased by milling to a lower flour extraction rate or only selecting the lowest ash streams from a straight grade flour. Refrigerated storage will also help to minimize color changes and prevent the increased deterioration that can occur when using sprouted wheat flour due to increased microbial or fungal contamination reflected in sliminess on the surface of the noodle.

10.7 Conclusions

Product deterioration caused by using sprout-damaged wheat is the result of one or several enzymes acting in concert during processing and/or prior damage to the storage reserves. Thus, characterization of the initial wheat prior to processing is important to ascertain the relative importance of these two types of damage. Certainly, reliance on one enzyme, α-amylase, as a measure of quality deterioration is less satisfactory than a more detailed examination.

There has been considerable research in the last two decades on enzyme systems present in sprouted wheat, such as the amylases and proteases. For example, we have considerable knowledge about their multiplicity and basic chemistry. We must now expand this knowledge to a better understanding of how these enzymes react with their natural substrates during the processing of various food products made from wheat flour or semolina. The relative importance of different enzymes will vary from product to product. Having a better understanding of the role of individual enzymes in degrading their substrate will assist in finding novel ameliorative remedies to prevent their disastrous effects. This is not an easy task for food scientists as the system is very complex. Furthermore, in the light of recent emphasis on natural foods, any addition of non-cereal inhibitors of enzymes is less likely to be received favorably. New milling processes[37,38] in which a substantial amount of the enzyme-rich bran is removed by pearling as part of the milling process may allow wheats with increased sprout damage to be used in processing in the future.

References

1. Derera, N. F. (1990) A perspective of sprouting research, in *Proceedings of the 5th International Symposium on Pre-harvest Sprouting in Cereals*, (ed. K. Ringlund, E. Mosleth and D. J. Mares) Westview Press, Boulder, p. 3.
2. Kruger, J. E. (1990) Nephelometric determination of cereal amylases. In *Proceedings of the 4th International Symposium on Pre-harvest Sprouting in Cereals*, (ed. D. J. Mares) Westview Press, Boulder, p. 584.
3. Hagberg, S. (1960) A rapid method for determining α-amylase activity. *Cereal Chem.*, **37**, 218.
4. Hagberg, S. (1961) Note on a simplified rapid method for determining α-amylase activity. *Cereal Chem.*, **41**, 202.
5. Perten, H. (1964) Application of the falling number method for evaluating α-amylase activity. *Cereal Chem.*, **41**, 127.
6. Kruger, J. E. (1981) Severity of sprouting as a factor influencing the distribution of α-amylase in pilot mill streams. *Can. J. Plant Sci.*, **61**, 817.
7. Jensen, Sv.A., Munck, L. and Kruger, J. E. (1984) A rapid fluorescence method for assessment of preharvest sprouting of cereal grains. *J. Cereal Sci.*, **2**, 187.
8. Kruger, J. E. (1985) Rapid analysis of changes in the molecular weight distribution of buffer-soluble proteins during germination of wheat. *Cereal Chem.*, **61**, 205–208.
9. Kruger, J. E. and Marchylo, B. A. (1985) Examination of the mobilization of storage

proteins of wheat kernels during germination by high-performance reversed-phase and gel permeation chromatography. *Cereal Chem.*, **62**, 1–5.

10. Kruger, J. E. and Marchylo, B. A. (1990) Analysis by reversed-phase high-performance liquid chromatography of changes in high molecular weight subunit composition of wheat storage proteins during germination. *Cereal Chem.*, **67**, 141–147.

11. Tkachuk, R. and Kruger, J. E. (1974) Wheat α-amylase. II. Physical characterization. *Cereal Chem.*, **51**, 508.

12. Kruger, J. E. (1972) Changes in the amylases of hard red spring wheat during growth and maturation. *Cereal Chem.*, **49**, 379.

13. Marchylo, B., LaCroix, L. J. and Kruger, J. E. (1980) α-Amylase isoenzymes in Canadian wheat cultivars during kernel growth and maturation. *Can. J. Plant Sci.*, **60**, 433.

14. MacGregor, A. W., Marchylo, B. A. and Kruger, J. E. (1988) Multiple α-amylase components in germinated cereal grains determined by isoelectric focusing and chromato-focusing. *Cereal Chem.*, **65**, 326.

15. Marchylo, B. A., Kruger, J. E. and Irvine, G. N. (1976) α-Amylase from immature hard red spring wheat. I. Purification and some chemical and physical properties. *Cereal Chem.*, **53**, 157.

16. Weselake, R. J., MacGregor, A. W. and Hill, R. D. (1983) An endogenous α-amylase inhibitor in barley kernels. *Plant Physiol.*, **72**, 809.

17. Weselake, R. J., MacGregor, A. W., Hill, R. D. and Duckworth, H. W. (1984) Purification and characteristics of an endogenous α-amylase inhibitor from barley kernels. *Plant Physiol.*, **73**, 1008.

18. Zawhistowska, U., Langstaff, J. and Bushuk, W. (1988) Improving effect of a natural α-amylase inhibitor on the baking quality of wheat flour containing malted barley flour. *J. Cereal Sci.*, **8**, 207.

19. D'Appolonia, B. A. (1983) 'Sprouted' flour coping with damage. *Bakers Digest, Milling Baking News*, **6**, 6.

20. Kulp, K., Roewe-Smith, P. and Lorenz, K. (1983) Preharvest sprouting of winter wheat. I. Rheological properties of flours and physicochemical characteristics of starches. *Cereal Chem.*, **60(5)**, 355.

21. Lorenz, K., Roewe-Smith, P., Kulp, K. and Bates, L. (1983) Preharvest sprouting of winter wheat. II. Amino acid composition and functionality of flour and flour fractions. *Cereal Chem.*, **60(5)**, 360.

22. Lukow, O. M. and Bushuk, W. (1984) Influence of germination of wheat quality. II. Modification on endosperm protein. *Cereal Chem.*, **61(4)**, 340.

23. Lukow, O. M. and Bushuk, W. (1984) Influence of germination of wheat quality. I. Functional (breadmaking) and biochemical properties. *Cereal Chem.*, **61(4)**, 336.

24. Ibrahim, Y. and D'Appolonia, B. L. (1979) Sprouting in hard red spring wheat. *Bakers Digest*, **53**, October, 17.

25. Kruger, J. E. and Marchylo, B. A. (1990) Quantitative analyses by RP-HPLC of storage protein breakdown in Canadian wheat upon germination, in *Proceedings of the 5th International Symposium on Pre-harvest Sprouting in Cereals*, (eds K. Ringlund, E. Mosleth and D. J. Mares) Westview Press, Boulder, p. 139.

26. Kruger, J. E. and Marchylo, B. A. (1990) Analysis by reversed-phase high-performance liquid chromatography of changes in high molecular weight subunit composition of wheat storage proteins during germination. *Cereal Chem.*, **67**, 141–147.

27. Finney, P. L., Morad, M. M., Patel, K., Chaudhry, S. M., Ghiasi, K., Ranhotra, G., Seitz, L. M. and Sebti, S. (1980) Nine international breads from sound and highly-field-sprouted Pacific Northwest soft white wheat. *Bakers Digest*, **6**, 22.

28. Orth, R. A. and Moss, H. J. (1987) The sensitivity of various products to sprouted wheat, in *Proceedings of the 4th International Symposium on Pre-harvest Sprouting in Cereals*, (ed. D. J. Mares) Westview Press, Boulder, p. 167.

29. Kruger, J. E. and Matsuo, R. R. (1982) Comparison of α-amylase and simple sugar levels in sound and germinated durum wheat during pasta processing and spaghetti cooking. *Cereal Chem.*, **51**, 26.

30. Dick, J. W., Walsh, D. E. and Gilles, K. A. (1974) The effect of sprouting on the quality of durum wheat. *Cereal Chem.*, **51**, 180.

31. Matsuo, R. R., Dexter, J. E. and MacGregor, A. W. (1982) Effect of sprout damage on durum wheat and spaghetti quality. *Cereal Chem.*, **59**, 468.

32. Combe, D., Garcon-Marchand, O., Seiller, M.-P. and Feillet, P. (1988) Influence de la germination sur la qualité des blés durs. *Ind. Céréales*, **53(3)**, 29.
33. Dexter, J. E., Matsuo, R. R. and Kruger, J. E. (1990) The spaghetti-making quality of commercial durum wheat samples with variable α-amylase activity, *Cereal Chem.*, **67(5)**, 405–412.
34. Bean, M. M., Keagy, P. M., Fullington, J. G., Jones, F. T. and Mecham, D. K. (1974) Dried Japanese noodles. I. Properties of laboratory-prepared noodle doughs from sound and damaged wheat flours. *Cereal Chem.*, **51**, 417.
35. Bean, M. M., Nimmo, C. C., Fullington, J. G., Keagy, P. M. and Mecham, D. K. (1974) Dried Japanese noodles. I. Effect of amylase, protease, salts, and pH on noodle doughs. *Cereal Chem.*, **51**, 427.
36. Kruger, J. E., Matsuo, R. R. and Preston, K. (1992) A comparison of methods for the prediction of Cantonese noodle color. *Can. J. Plant Sci.*, **72**, 1021.
37. Sugden, D. (1992) Tkac and Timm process: a miller's perspective. *World Grain*, **9(4)**, 16.
38. Reeves, W. H. (1991) New technology for the optimization of flour milling. *Cereal Foods World*, 681.

11 Soft wheat quality in production of cookies and crackers

H. FARIDI, C. GAINES and P. FINNEY

11.1 Introduction

Of the five principal classes of wheat grown in the United States, soft red winter, soft white winter and spring, as well as soft white winter club are the wheats mainly used in production of cookies and crackers (Figure 11.1). The terms 'hard' and 'soft' as applied to wheats are descriptions of the texture of the wheat kernel. Flour obtained from a hard wheat kernel has a coarser particle size than does flour obtained from a soft wheat kernel. The number of types of products made from soft wheat is large. A partial list is shown in Table 11.1. All of the listed products have better appearance and eating quality when made from soft rather than hard wheat flour. Low-protein (7–10%) flours milled from soft wheats are most suitable for making cakes and biscuits.[1]

Soft wheats differ from hard wheats in kernel hardness, a basic genetic directly inherited characteristic. When ground or milled, soft wheat

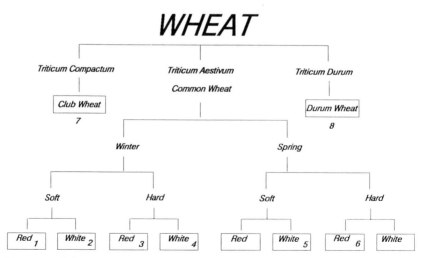

Figure 11.1 US wheat classes. Major uses: 1, 2, 5, 7, cookies, crackers and cakes; 3, 4, 6, various types of breads; 8, pasta.

Table 11.1 Products made from soft wheat[a]

Biscuits	Pancakes
Cookies	Doughnuts
Crackers	Oriental noodles
Wafers	Thickening agent for soups and soup mixes
Pretzels	Crumbs for coating fish and meat products
Cakes	Breakfast cereals
Pastry products from pie crust to sweet Danish pastry	Flat breads
Waffles	Ice cream cones

[a] From Faridi, ref. 3, with permission.

generally fractures into significantly smaller particles than does hard wheat, which is reflected by its greater 'break flour yield' upon milling. Break flour is the portion of the kernel endosperm obtained without crushing or reduction in the milling operation. Soft wheat cultivars exhibit a wide range in kernel texture, but historically there has been a distinct difference in average flour particle size between the hardest soft wheats and softest hard wheats, such that a relatively simple grinding and sieving test can properly classify wheat.

To achieve greater baked product uniformity and consistency, specifications are adopted by millers and bakers which vary between companies as well as from one baking plant to another. However, such specifications all have some common features. Modernization and automation of bakeries has necessitated exact and demanding specifications on incoming flour. The following are general specifications for soft wheat and its flour, which vary according to the particular end-product.[2]

Wheat:
- High test weight or 100-kernel weight; uniformity of kernel size,
- Ease of milling and high flour yield,
- Low to medium protein content (8–11%, on 14% moisture basis),
- Moisture content not exceeding 13%,
- Kernel softness (particle size index PSI > 55s),
- Low amylase activity; no sprout damage.

Flour:
- Bright and creamy color and relatively low ash content,
- Low to medium protein content (7–10%, on 14% moisture basis),
- Low water absorption,
- Low damaged-starch content,
- Fine flour granulation,
- Medium mixing requirements and satisfactory dough-handling properties,
- Tender-bite cookies and crackers.

Table 11.2 Characteristics of flour used for cookies/crackers production[d]

Product type	Wheat[a] blend	Typical ranges			
		Protein (%)	Ash (%)	Viscosity[b] (M)	Spread factor (w/t)[c]
Cookies					
Wire cut	SWW–SRW	7.0– 8.0	0.38–0.42	20–40	8.5–10.0
Rotary	SWW–SRW	8.0– 9.0	0.38–0.42	30–50	7.5– 9.0
Deposit	SWW–SRW	7.5– 8.5	0.38–0.42	25–45	8.0– 9.5
Crackers					
Soda-sponge	SRW–HRW	9.0–10.0	0.41–0.45	55–70	6.5– 8.0
Soda-dough	SRW	8.5– 9.5	0.39–0.43	40–55	7.5– 9.0
Snack	SRW	8.5– 9.5	0.40–0.44	40–55	7.5– 9.0

[a] SRW = soft red winter, SWW = soft white winter; HRW = hard red winter, [b] MacMichael value. [c] Width/thickness. [d] From Tanilli, ref. 4, with permission.

Flour constitutes the primary raw material to which all soft wheat product formulations are related. It provides a matrix into which other ingredients in varying proportions are mixed to form dough or batter systems.[3,4]

There are so many different types of baked products that no one criterion of flour can satisfy all requirements of the numerous soft wheat products. Key factors pertinent in characterizing soft wheat flours are significant measurable parameters and operational experience. In this chapter, only those qualities of soft wheat flour will be discussed that are related to its functionality in cookie and cracker production.

Cookie and cracker flours (Table 11.2) receive no special treatment or additives, i.e. they are not normally bleached or chemically matured in any way and have no chemical leavening additives such as phosphates or self-raising ingredients. At times, cookie flours are very lightly chlorinated to reduce and control the 'spread' of a product.

11.2 Evaluation of soft wheat flour

Most physical methods used to evaluate wheat flour were developed to measure hard wheat flour properties used in production of bread doughs. However, most products made from soft wheat flours do not require doughs that are mixed or developed to the extent that bread doughs are. Therefore, it would appear illogical to use techniques designed for hard wheat flours to evaluate soft wheat flours.[5] The use of physical dough-testing machines such as the mixograph, farinograph, or alveograph to evaluate soft wheat flours presumes that the rheological properties of soft wheat gluten are opposite to those of hard wheat gluten. Stated another way, if a good hard wheat flour gives a strong gluten network, then a good soft wheat flour should give a weak gluten network. There appears to be little practical evidence to support that assumption.

The physicochemical properties of a good quality soft wheat flour constitute a unique blend of properties. They are not the properties of a poor quality hard wheat flour.[5] However, this situation does not negate the value of physical dough-testing equipment in soft wheat flour testing and evaluation. As for hard wheat flour, such instruments demonstrate the variability or uniformity of gluten strength between different lots of flour. Since flour is the major ingredient in soft wheat products, its uniformity is of extreme value to the commercial baker or end-user. Having said that, the rheological properties of flour are of only limited commercial value in predicting the biscuit-making properties of soft wheat flours.[5-7]

Starch damage is another important quality factor for soft wheat flours, and thus its measurement is considered important. A high level of starch damage is undoubtedly detrimental to cookie making, because it is invariably linked with wheat hardness and elevated protein and milling damage. Reported correlations between starch damage and cookie diameter are, however, poor.[8] Damaged starch absorbs much more water than does undamaged starch. The amount of water absorbed by a flour is an important quality of cookie flours; one that absorbs small amounts of water is desired. Gaines *et al.* demonstrated that elevated damaged starch increases sugar-snap cookie dough stiffness and consistency and decreases cookie diameter.[9]

Another important quality of soft wheat flours is small particle size.[5,10,11] This appears not to be very important in cookie flours, perhaps because the production of fine particles also results in a high level of starch damage. The net result is no improvement in cookie quality if particle size is overly reduced by milling. However, cultivars that produce larger cookies and cakes tend to be those that produce flour of smaller particles size, perhaps because such flours have less damaged starch.[12]

One of the most useful tests to evaluate soft wheat flours is the alkaline water retention capacity (AWRC) test.[5,13] This relatively simple procedure[14] is reasonably good for assessing soft wheat flours and roughly predicts flour performance in cookie baking in the laboratory, using the official AACC[15] sugar-snap cookie method. The ability of a flour to hold water against centrifugal force is a measure of a flour's affinity for water. Good quality soft wheat cookie flours hold water poorly.[5]

11.3 Factors affecting soft wheat quality

Average wheat production per unit of land has doubled in the last two decades.[16] The introduction and popularity of high-yielding, semi-dwarf cultivars have been the most important factors causing this increase in production. Wheat breeding programs are in a competitive struggle between the need for increased field yields and the need for wheat with desired milling and baking characteristics. Wheat breeders, whose long-term priorities are to maximize all agronomic and wheat quality characteristics, are increasingly forced to make choices between field yield and end-use quality when selecting for release of new cultivars. However, hard pressed farming economics and declining wheat prices necessitate the choice of yield as the major criterion in wheat breeding. That is not to say that milling and baking qualities are not considered in most wheat breeding programs.

Heritable differences among wheat cultivars are only part of the reason why wheat milling and flour baking qualities vary within and between crop years. Wheat is grown in all 48 contiguous states of the USA and is therefore subjected to enormous variety in growing conditions. Kernel hardness, and therefore bread flour yield, is also influenced by the time and extent of fertilizer application, environmental conditions, and different growing seasons.[17,18]

To illustrate the effects of growing conditions on wheat and flour quality, a frequency distribution representing 35 determinations of break flour yield of Caldwell soft red winters (SRW) wheat, grown in the midwest US, is presented in Figure 11.2. Those variations in wheat hardness due to growing conditions can cause considerable variations in milling or baking quality.[18]

Individual farming practices also affect wheat and flour quality. Planting time can alter disease susceptibility, which in turn affects wheat quality. If wheat is harvested at too high a moisture content, incipient seed germination can result and be detrimental to the quality of some cookies and crackers.

11.4   Critical factors in soft wheat quality for commercial bakers

In the past two decades, commercial bakers in the United States have developed stringent flour specifications based on flour composition (e.g. moisture, protein, ash contents), as well as functionality (e.g. mixograph, farinograph, and alveograph) tests. These specifications have helped the baking industry to improve the uniformity of incoming flour. However, it is still unclear how the physical and chemical attributes of the flour have an impact on large-scale production of baked products.

Three major questions that concern cookie and cracker manufacturers are the impact of flour protein content, gluten strength, and wheat kernel hardness on product textural attributes. Soft wheats that have hard kernels present problems for cookie and cracker manufacturers. Cookies and crackers are very dense compared to bread and rolls. Hard kernel characteristics are in some way transmitted to the product, resulting in an undesirable hard or 'flinty' texture. Traditionally, kernel hardness of wheat

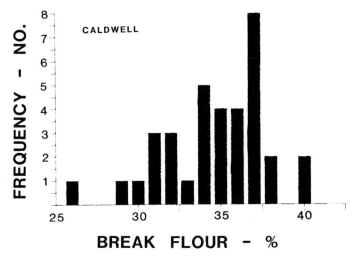

Figure 11.2 Frequency distribution of break flour yields for 35 determinations of Caldwell wheat grown in several locations during several crop years. From AACC, ref. 14, with permission.

was considered to be the miller's problem, because hardness directly affected such factors as the energy requirement for milling and the roll settings during milling. However, kernel hardness also influences the extent of damaged starch produced in a flour during the milling process. Other relationships between kernel hardness and product hardness may also be important.

In cookie and cracker manufacturing, increasing the levels of sugar and fat in a dough reduces the 'flinty' nature of a product, when harder wheat flours are used. Therefore, the texture of products high in sugar and fat, such as sugar-snap cookies, is not affected as much as is the texture of lower sugar, lower fat products, such as snack or saltine crackers.[17]

The hardness of a given variety of wheat is genetically controlled and is not directly related to the protein content of the grain.[19–21] The amount of protein present in a biscuit flour is important, as it affects a number of properties, including texture of cracker and cookie doughs and their products.[6,22–24] In order to maintain a satisfactory inner structure in a cracker, Wade[24] found it necessary to adjust the final gauge-roll gap, so as to maintain approximately constant product weight. The relationship observed between gauge-roll gap and flour protein content is shown in Figure 11.3. The hardness of semi-sweet biscuits, as measured by a texture meter, increases with increasing flour protein content[22] (Figure 11.4). The subjective appearance of semi-sweet biscuits was found to be closely related to the protein content of the flour used in their manufacture.[24]

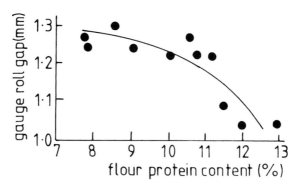

Figure 11.3 Relationship between the gauge-roll gap required to produce cream crackers of constant weight and the protein content of the flour used in their manufacture. From Finney *et al.*, ref. 18 and Simmonds, ref. 20, with permission.

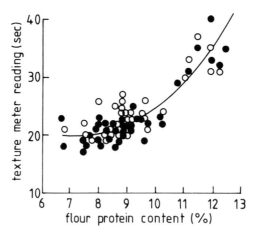

Figure 11.4 Effect of flour protein content on the texture meter readings of semi-sweet biscuits. Unsulphited doughs, ●; sulphited doughs, ○. From Wade, ref. 6 and Simmonds, ref. 20, with permission.

11.5   Textural properties of cookies and crackers

Measuring the texture of cookies and crackers is a difficult task, complicated by their unique and heterogeneous nature and inconstant structure. Cookie and cracker geometry causes heat to penetrate unevenly during baking creating longer residence of moisture in the center of a product and allowing greater amounts of starch to gelatinize there, which also results in more open space (crumb structure) in the center.[25,26]

Cookies and crackers usually have low-moisture content and a wide range of hardness, brittleness, and crispness. Product moisture content has a differing and significant influence on various aspects of the measured texture of cookies and crackers. Cookies and crackers are sometimes docked with pins to create holes or to force together laminated structures, in order to create more uniform baking, texture, and appearance. They may also be sprayed with oil, sometimes unevenly, which may influence measurement.[24] Product texture also will be affected if the texture of a particular product is a function of its geometry. As we are all only too aware, crop year and cultivar of wheat cause significant variance in the geometry of products baked from wheat flour.

The best agreement between sensory and instrumental texture data often occurs using principles of instrumental measurement most similar to those used during sensory assessment. For good correlations between sensory and instrumental measurements, both measurements should be conducted using similar intensities of applied force appropriate for a given food item.

Gaines et al.[25,27–30] recently reported studies relating genetic and environmental factors to cookie texture, using the AACC sugar-snap cookie, as well as a wire-cut cookie formula, more typical of commercial products, developed by Slade and Levine[31] (Table 11.3).

As Figure 11.5 shows, the hardness of the AACC sugar-snap cookie is not generally associated with flour protein content, whereas the wire-cut formula cookies do show this relationship as also do commercial products. As explained by Slade and Levine,[31] the higher concentration of sucrose in

Table 11.3 Comparison of formulations[a]

Ingredient	AACC 10–52		Model wire-cut	
Flour	100	14% (moisture basis)	100	13% (moisture basis)
Sugar	60	bakers' special	42	32 fine granulated
Brown sugar				10
Shortening	30		40	
High fructose corn syrup (HFCS)			1.5	
Non-fat dry milk (NFDM)	3		1	
Salt	0.45		1.25	
Sodium bicarbonate	1.8		1	
Ammonium bicarbonate			0.5	
Ammonium chloride	0.5			
Water	15	vary to adjust rheology	22	constant
Total				
Sugar + water	75		64	
Ratio				
Sugar/water	4		1.9	
Sugar particle size	177 μ		297 μ	

[a] Extracted from ref. 31, courtesy of L. Slade and H. Levine.

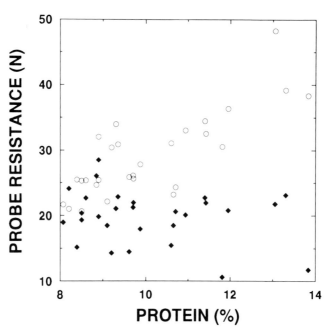

Figure 11.5 Effect of protein content of 28 flour samples on textural hardness of cookies made with AACC sugar-snap cookie procedure $r = -0.22$ (◆); or commercially based wire-cut formulation $r = 0.79$ (○). From Wade, ref. 24, with permission.

the sugar-snap cookie formula cools after baking to form an aqueous glass. The sucrose–water glass gives the cookie a relatively continuous hard structure that is largely unaffected by other potential contributors to cookie hardness. The wire-cut formulation, being more like commercial products, has less sucrose. Wire-cut type cookies do not go through a sucrose–water glass transition during cooling after baking and thus remain more sensitive to other textural operatives. As one would expect, the result is that the hardnesses of the two cookie formulations do not correlate (Figure 11.5).

When a single cultivar, Caldwell (grown at one location in Ohio and having a range of flour protein), was tested using the bench-top wire-cut cookie formula, the effect of increasing flour protein content on cookie texture became more evident[28] (Figure 11.6).

Gaines *et al.*[28] also studied the role of flour components on cookie textural hardness by fractionation–reconstitution of two Ohio soft red wheat cultivars that produce harder or softer wire-cut formula cookies. Contributions of each fraction to hardness of wire-cut formula cookies increased in the order: starch, lipid, tailings, gluten, water solubles, gluten + tailings, gluten + water solubles, water solubles + tailings and gluten + water solubles + tailings (Figure 11.7). The relative effectiveness of interchanging combinations of two or three fractions was generally predictable from the

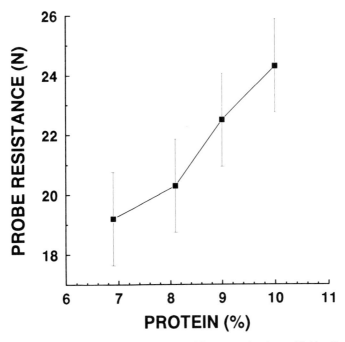

Figure 11.6 Effect of flour protein content on cookie texture hardness (Caldwell cultivar). From Wade, ref. 24, with permission.

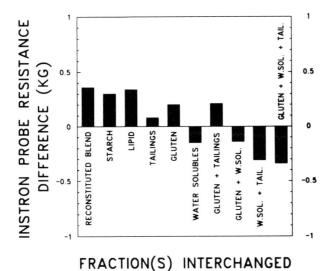

Figure 11.7 Difference between the texture of cookies made from fractions interchanged from two cultivars that produced hard and soft cookies. From Wade, ref. 24, with permission.

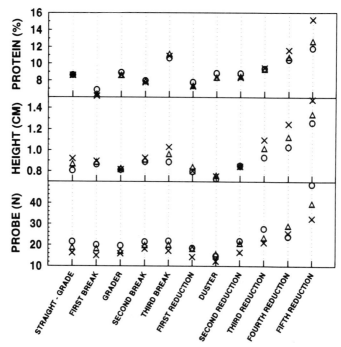

Figure 11.8 Cookie hardness values produced from straight-grade flour and ten millstreams produced using a Miag Multomat mill. ○ = soft white winter club wheat; △ = commercial mill blend of wheats; × = soft red winter wheat.

effectiveness of interchanging individual fractions. Differences in hardness were completely reversed when the three most individually effective fractions (water solubles, tailings, and gluten) were interchanged.

Millstreams are well known to have different baking qualities, with streams produced earlier in the mill flow being lower in protein and damaged starch. After observing that elevated flour protein content produced smaller, thicker, and harder wire-cut formula cookies, Gaines *et al.* made wire-cut cookies from ten millstreams produced by a Miag Multomat automated laboratory mill.[30] Three wheats were milled: a soft white winter club, a commercial mill blend of wheats, and a soft red winter (Figure 11.8). Flour protein content and cookie height increased as milling flow continued. For each wheat, cookie hardness (probing resistance) was relatively similar among the first, second, and third breaks; first and second reductions; and the grader stream. Thereafter in the mill flow, cookie hardness increased in the order: third < fourth < fifth reductions. Except for one millstream, differences in cookie hardness resulting from the three wheats were consistent within millstreams. The club wheat produced harder cookies and SRW wheat softer cookies, when produced from nine of the ten millstreams. This finding suggests that it would be difficult to use millstream selection to

Table 11.4 Cultivar softness and protein quality[a]

	Break flour yield		Mixograph number		Protein	
	1988 Crop (%)	1989 Crop (%)	1988 Crop	1989 Crop	1988 Crop (%)	1989 Crop (%)
Compton	25.3	31.6	87.7	126.0	8.6	10.0
Caldwell	30.4	39.2	123.0	89.0	9.3	7.7
Becker	28.6	39.2	60.6	27.9	10.1	6.9
Hillsdale	25.8	34.6	59.0	39.5	10.8	8.3
Average	27.5	36.2	82.6	70.6	9.7	8.2

[a] From Gaines et al., ref. 30, with permission.

produce cookies, from a wheat that produces relatively hard cookies, to be as tender as those produced from a wheat that produces comparatively soft cookies.

Another study was conducted by members of the Flour Committee of the Biscuit and Cracker Manufacturers Association (B&CMA).[32] The participating companies (Nabisco Biscuit Company, Keebler, Sunshine, Archway Cookies, Balsans, Schulze and Burch, and Interbake) collaborated with the staff of USDA-ARS Soft Wheat Quality Laboratory in Wooster, Ohio, and Mennel Milling Company to conduct a two-year study to determine the effects of flour protein quality and wheat hardness on cookie and cracker texture.

Samples of four pure cultivars (Compton, Hillsdale, Becker, and Caldwell), all midwest US SRW wheats, were milled at Mennel Milling Company in Fostoria, Ohio. The test was conducted twice, using 1988 and 1989 grown cultivars. These cultivars were chosen to provide reasonable ranges of gluten quality (as measured by mixograph and reported as mixograph number, and wheat hardness as measured by break flour yield) (Table 11.4). Allotments of the four flours were shipped to the member companies for pilot plant or plant trials. Various types of cookies (rotary-molded or wire-cut), as well as chemically-leavened crackers, were produced. The products were submitted for sensory testing and objective texture measurements to the USDA Soft Wheat Laboratory.

In addition to sensory evaluation and Instron testing of the above samples, the USDA laboratory staff thoroughly evaluated the flours and conducted bench-top cookie baking, using the previously mentioned wire-cut cookie formula. The data shown in Table 11.4 clearly corroborated industry reports that, on average, the 1989 crop was softer (break flour yield), weaker (gluten strength), and lower in protein than the 1988 crop. The hardness of bench-top cookies, as measured by Instron probing, was significantly greater for the 1988 crop flours (Table 11.5). Likewise, the cookies from the 1988 trials were significantly harder than those made from

Table 11.5 Textural hardness/tenderness of cookies made with B&CMA flours[a]

	Instron force (N)	
	1988 Crop	1989 Crop
Compton	22.5	23.1
Caldwell	24.4	15.9
Becker	29.0	22.1
Hillsdale	22.1	18.8
Average	24.5	19.97

[a] From Gaines *et al.*, ref. 30, with permission.

the 1989 crop flours. Since the 1988 crop wheat cultivars were genetically identical to those from the 1989 crop, the above findings indicated that environmentally induced changes in protein content, gluten strength, and kernel hardness will affect textural properties of cookies and crackers.

The overall sensory properties of all the cookies and crackers made at the pilot plant and plant facilities of the participating companies, for both 1988 and 1989 crop flours, are shown in Table 11.6. Within each crop year, the cultivars were grouped as soft or hard, and as strong, medium, or weak gluten mixing strength. All sensory data collected from all the baking companies, on all the cookies and crackers, were pooled and averaged. Therefore, as a generalization that can only be considered a trend, the data demonstrate that between and within each crop year, both flour protein strength and kernel texture appeared to have an impact on the textural attributes of cookies and crackers. Cookies and crackers made from softer and weaker gluten flours tended to have a more tender and more desirable (to most consumers) texture or 'bite' than did products made from stronger and harder wheat flours.

Table 11.6 Pooled sensory values of cookies and crackers made from B&CMA flours[a]

Wheat attributes		1988		1989		Combined	
Protein quality	Strong	2.28		3.48		2.88	
	Medium	2.10		1.83		1.96	(more tender)
	Weak	2.80		2.35		2.57	
Hardness	Fine	2.63		2.14	(more tender)	2.39	(more tender)
	Coarse	2.37	(more tender)	2.86		2.62	

[a] Higher values refer to harder products. From Gaines *et al.*, ref. 30, with permission.

References

1. Hoseney, R. C. (1986) *Principles of Cereal Science and Technology*, American Association of Cereal Chemists, St. Paul Minnesota, pp. 245–276.
2. Faridi, H. and Finley, J. W. (1989) Improved wheat for baking, *CRC Crit. Rev. Food Nutr.*, **28**, 175–209.
3. Faridi, H. (1990) Soft wheat products *Handbook of Cereal Science and Technology*, (ed. K. Lorenz and K. Kulp) Marcel Dekker, New York, pp. 683–739.
4. Tanilli, V. H. (1976) Characteristics of wheat and flour for cookie and cracker production. *Cereal Foods World*, **21**, 624–628, 644.
5. Hoseney, R. C., Wade, P. and Finley, J. W. (1988) Soft wheat products. In *Wheat Chemistry and Technology*, 3rd edn, (ed. Y. Pomeranz) American Association of Cereal Chemists, St. Paul, Minnesota, pp. 407–456.
6. Wade, P. (1972) Flour properties and the manufacture of semi-sweet biscuits. *J. Sci. Food Agric.*, **23**, 737–744.
7. Wainright, A. R., Crowley, K. M., and Wade, P. (1985) Biscuit making properties of flours from hard and soft milling single variety wheats. *J. Sci. Food Agric.*, **36**, 661–668.
8. Abboud, A. M., Rubenthaler, G. L. and Hoseney, R. C. (1985) Effect of fat and sugar in sugar-snap cookies and evaluation of tests to measure cookie flour quality. *Cereal Chem.*, **62**, 124–129.
9. Gaines, C. S., Donelson, J. R. and Finney, P. L. (1988) Effects of damaged starch, chlorine gas, flour particle size, and dough holding time and temperature on cookie dough handling properties and cookie size. *Cereal Chem.*, **65**, 384.
10. Yamazaki, W. T. (1959) Flour granularity and cookie quality. I. Effect of wheat variety on sieve fraction properties. *Cereal Chem.*, **36**, 42–51.
11. Yamazaki, W. T. (1959) Flour granularity and cookie quality. II. Effects of changes in granularity on cookie characteristics. *Cereal Chem.*, **36**, 52–59.
12. Gaines, C. S. (1985) Associations among soft wheat flour particle size, protein content, chlorine response, kernel hardness, milling quality, white layer cake volume, and sugar-snap cookie spread. *Cereal Chem.*, **62**, 290.
13. Yamazaki, W. T. (1953) An alkaline water retention capacity test for the evaluation of cookie baking potentialities of soft winter wheat flours. *Cereal Chem.*, **30**, 242–246.
14. AACC (1983) *Approved Methods of the American Association of Cereal Chemists*, 8th edn, Method 56–10, The Association, St. Paul, Minnesota.
15. AACC (1983) *Approved Methods of the American Association of Cereal Chemists*, 8th edn, Method 10–50D, Method 10–52, The Association, St. Paul, Minnesota.
16. Briggle, L. W. and Curtis, B. C. (1987) Wheat worldwide. In *Wheat and Wheat Improvements* (ed. E. G. Heyne) American Society of Agronomy, Madison, Chap. 1.
17. Faridi, H. A., Finley, J. W. and Leveille, G. A. (1987) Wheat hardness: a user's view. *Cereal Food World*, **32**, 327–329.
18. Finney, P. L., Gaines, C. S. and Andrews, L. C. (1987) Wheat quality: a quality assessor's view. *Cereal Foods World*, **32**, 313–318.
19. Yamazaki, W. T., and Donelson, J. R. (1983) Kernel hardness of some US wheats. *Cereal Chem.*, **60**, 344–350.
20. Simmonds, D. H. (1974) Chemical basis of hardness and vitreosity in the wheat kernel. *Bakers Digest*, **48**(5), 16–18, 20, 22, 24, 26–29, 63.
21. Miller, B. S., Pomeranz, Y., and Afework, S. (1984) Hardness (texture) of hard red winter wheat grown in a soft wheat area and of soft red winter wheat grown in a hard wheat area. *Cereal Chem.*, **61**, 201–203.
22. Wade, P. (1988) *Biscuit, Cookies and Crackers, Vol. 1, Principles of the Craft*, Elsevier Applied Science, London.
23. Wade, P. (1970) Technological aspects of the use of sodium metabisulphite in the manufacture of semi-sweet biscuits, *Food Trade Rev.*, **40**(5), 34–39.
24. Wade, P. (1972) Flour properties and the manufacture of cream crackers. *J. Sci. Food Agric.*, **23**, 1221–1228.
25. Gaines, C. S. (1991) Objective measurements of the hardness of cookies and crackers. *Cereal Foods World*, **36**, 989.
26. Burt, D. J. and Fearn, T. (1983) A quantitative study of biscuit microstructure. *Starch*, **35**, 351–353.

27. Gaines, C. S., Kassuba, A. and Finney P. L. (1992) Instrumental measurement of cookie hardness. I. Methods assessments. *Cereal Chem.*, **69**, 115.
28. Gaines, C. S., Kassuba, A., Finney, P. L. and Donelson, J. R. (1992) Instrumental measurement of cookie hardness. II. Application to product quality variables. *Cereal Chem.*, **69**, 120.
29. Gaines, C. S., Kassuba, A., and Finney, P. L. Effect of kernel softness, protein content, and milling extraction stream on wire-cut formula cookies. *Cereal Chem.*, submitted.
30. Gains, C. S., Kassuba, A., and Finney, P. L. Instrumental measurement of cookie hardness. IV. Miag Multomat Millstream Selection, *Cereal Chem.*, submitted.
31. Slade, L. and Levine, H. (1993) Structure–function relationships of cookie and cracker ingredients. In *The Science of Cookie and Cracker Production*, (ed. H. Faridi) Van Nostrand Reinhold/AVI, New York, in press.
32. Faridi, H. (1990) *Report on the Two-year Study on Soft Wheat Quality Attributes affecting Cookie and Cracker Texture*. Presented at the B&CMA Annual Technical Conference, October 28–30, Orlando, available through B&CMA Office, Washington, DC.

12 Durum wheat: its unique pasta-making properties
R. R. MATSUO

12.1 Introduction

Wheat is the major food crop in the world, grown in most countries except in the hot, humid tropical regions. World production of wheat in the years 1988–1990 averaged 548 million tonnes per year.[1] Most of this production, well over 90%, is common wheat, *Triticum aestivum*. This species encompasses hundreds of varieties with a wide range of quality characteristics. The seed coat can be red or white, the kernel hardness ranges from very soft to hard, the protein content ranges from 6.5 to 20%, and varieties can be spring or winter habit. From such a wide range of intrinsic quality it is not surprising that a myriad of flour products are produced and consumed throughout the world.

Durum wheat, *Triticum turgidum* var. durum, accounts for about 5% of the total wheat production. Unlike common wheat, there is only one predominant class of durum wheat. Most varieties are spring or semi-winter types. Only a few true winter habit durums are known. Kernels are amber in color, large with kernel weight ranging from 35 to 60 mg and very hard.

Durum wheat is similar in composition to common wheat. The percentages of starch, protein, minerals, lipids and amino acids are roughly equivalent. One of the notable differences is the mineral distribution in the kernel. This difference may contribute to the superior cooking quality of pasta from durum semolina. Results of cooking tests of spaghetti processed from the two classes of wheat indicate the importance of minerals both in the cooking water and in wheat.

Other species of wheat are grown but represent only a very small percentage of total wheat production. In California and the Pacific northwest of the USA, club wheat, *Triticum compactum* is grown. In India, emmer, *Triticum dicoccum*, accounts for about 1% of the wheat cultivated area.[2]

12.2 Durum wheat production area

Major areas of production are given in Table 12.1. Near East Asia includes Turkey, the largest producer accounting for about 20% of the world durum

Table 12.1 Estimated area, yield and production in durum producing regions[3]

Region	Area (1000 ha)	Yield (tonnes ha^{-1})	Production (1000 tonnes)
Western Europe	2490	2.30	5730
North America	2960	1.94	5756
South America	102	1.92	196
Near East Asia	4462	1.56	6950
North Africa	3290	0.98	3214
Others	3756	0.94	3540
World total	17060	1.49	25360

production.[1] In western Europe the major producers are Italy, France and Greece. In the past several years Canada has been second to Turkey in production. North African countries producing durum wheats are Morocco, Algeria and Tunisia. 'Others' include the former USSR, Eastern Europe and Far East Asia.

In the first half of this century Russia was the largest producer and exporter of durum wheat.[4] Currently, figures are unavailable for production in Russia. Another country where durum is grown but figures are not readily available is India. According to Pandy,[2] durum wheat accounts for 14% of the total cultivated wheat area there.

Durum wheat is better suited than common wheat to areas where the annual precipitation is low (300–450 mm). Good yields can be obtained under irrigation but this practice is limited to very few areas. The so-called desert durums are grown under irrigation in Arizona, New Mexico and California.

High summer temperatures and low humidity improve grain quality. Durum wheat thrives best where a considerable proportion of the annual rainfall occurs during the vegetative phase and when occasional showers alternate with sunshine and hot dry conditions during the grain filling period.[5] Rain during the harvest season can cause problems as durum wheats are much more susceptible to sprouting than common wheats.

12.3 Uses of durum

Although durum wheat as a class might be considered a minor crop relative to common wheat, the diets of millions of people in the Middle East and North Africa are based on durum wheat. In Western Europe and North America durum wheat is consumed primarily in the form of pasta products. On the other hand, in the Near East and North Africa durum wheat is consumed in various products as shown on Table 12.2.

Table 12.2 Principal use of durum wheat in the Near East and North Africa[6]

Durum wheat products	%
Two-layer breads	35
Single layer	18
Burghul	15
Cous-cous	10
Pasta	15
Frekeh	2
Feed	5

Two-layered breads are known by various names in different countries; khobz, baladi and shami. Single-layered breads include tannour, saaj, mountain bread, markouk and mehrahrah.[6] Burghul (or sometimes spelled bulgur, boulgur, borghol, boulghour) is granulated, boiled, sun-dried wheat used in several ways, ranging from steamed burghul much like rice, to burghul/minced meat dishes and sweets. Frekeh is green, dried, parched wheat ground into coarse particle size and steamed like rice.

In Sicily durum is used for leavened bread. This apparently is also a common practice in Greece[7] and popularly known as peasant bread. Trahana is another use of coarsely ground wheat mixed with milk and used in soups. It is well known in Balkan countries.[7] Semolina cakes are another form of consumption in the Near East. They are rich in butter and soaked in sugar syrup.

Quality factors considered desirable for some of the above products have been listed by Williams *et al.*[8] as given in Table 12.3.

Quality requirements for these products are very similar to those for pasta. In North America and Western Europe where research on durum wheat is conducted, the focus is on pasta quality. Thus, the vast majority of papers on durum wheat are concerned with requirements for pasta.

Table 12.3 Desirable quality factors for flat breads, cous-cous and burghul

Product	Desirable quality factors
Flat breads	Medium protein, high gluten strength, light color
Cous-cous	High vitreous kernel count, high protein, high gluten strength, high semolina yield, yellow color
Burghul	High vitreous kernel count, high protein, high gluten strength, high semolina yield, yellow color

12.4 Durum wheat quality

There is very little difference between the chemical composition of durum and common wheat. The percentage of protein, carbohydrates, lipids and mineral matter is very similar. Amino acid composition of gluten protein too is very similar between the two classes. Yet it is generally agreed that durum wheat is the preferred raw material for the production of pasta products, although for high volume pan bread or for baguettes of acceptable quality, common wheat is far superior.

The question of what the basic differences are between common and durum wheats has not been resolved. We know there are differences in the physical characteristics of freshly washed-out gluten between the two classes. If one were to compare their characteristics, differences are readily detectable. The SDS sedimentation volume[9] (the Zeleny sedimentation test modified by addition of sodium dodecyl sulfate), a measure of gluten strength, is much higher for bread wheat than for durum wheat. Dexter et al.[10] reported a mean SDS sedimentation volume of 41.8 ml for 22 samples of durum wheat and 75.1 ml for 38 samples of bread wheat. Gluten strength is also reflected in normal farinograms, i.e. mixing curves titrated to 500 BU. Dough development time is shorter for durum wheat (3.27 min versus 5.42 min)[10] and stability is shorter.

Dough strength, in terms of farinograph characteristics, is reflected in baking quality, which in turn can be represented by the baking strength index (BSI),[11] a protein quality parameter that expresses loaf volume by the 'Remix' baking method, as a percentage of the volume normally expected for Canadian hard red spring wheat flour of the same protein content. BSI for durum varieties averaged 75.1% while that for red spring averaged 105.4%.[10]

In spite of the fact that gluten strength is stressed as essential for good cooking quality in spaghetti,[12] strength in hard red spring wheat does not produce top quality pasta. Some studies have shown the pasta made from hard red spring farina was better than that made from durum semolina,[13,14] while others have shown the reverse.[10,15] A recent study by Malcolmson et al.[16] again showed that the cooking quality of durum semolina spaghetti was superior to that of spring wheat farina pasta.

12.5 Mineral constituents

The results of an international collaborative test on durum wheat quality indicated that on the basis of multiple regression analysis of wheat data, wheat ash was the major single factor determining firmness of cooked spaghetti.[17] Semolina ash did not appear to exert the same influence. This was somewhat difficult to rationalize since spaghetti is processed from semolina, not from wholewheat. However, the notion of mineral content influencing cooking quality has been stated by an Italian pasta manufacturer (R. R. Matsuo, 1991, personal communication). He claimed that the best pasta made in Italy could not be marketed because it exceeded the legal limit of ash content, in excess of 0.90% ash. This product was being produced from high ash high extraction semolina. In laboratory tests this sample was very firm with resilience.

Mineral constituents in wheat account for about 1.60% (range 1.25–2.40%) of the wheat kernel. This amount is approximately the same for common bread and durum wheats. The distribution of mineral matter within the kernel, however, differs significantly between the two species.

Straight grade flour milled to 75% extraction from hard red spring wheat normally contains about 0.50% ash. Durum wheat flour, on the other hand, milled to the same extraction contains about 0.75–0.80% ash. A more striking difference is in the ash content of gluten, about 0.25% in the flour of red spring wheat and about 1.0% in that of durum wheat.

There appears to be some basis for considering the mineral constituent as a factor in pasta cooking quality. Four elements predominate in both durum semolina and hard red spring flour; potassium, magnesium, calcium and phosphorus.[18] Durum semolina has a higher content of potassium, magnesium and phosphorus than hard red spring flour as shown on Table 12.4.

Durum semolina contains 1.7 times more potassium, 2.0 times more magnesium and close to 1.5 times more phosphorus than red spring flour. What functional role potassium, the mineral constituent in the highest concentration, plays has not been elucidated. More studies have been carried out on phosphorus in phytic acid (inositol hexaphosphoric acid) and in phospholipids. Phytic acid is the main phosphorus-containing compound in wheat, accounting for 72–78% of the total phosphorus in hard wheats and about 66% in durum wheats.[18] In durum semolina phytic acid comprises only 2.7–6.0% of the total phytate in wheat,[19] while in hard wheat flour of the percentage ranges from 15.1 to 45.5%.[18] Phospholipids or phosphoglycerides are the other important phosphorus-containing compounds. Their association with gluten is well established,[20] and have been implicated in baking quality.

Table 12.4[a] Mineral content in durum semolina and hard red spring flour

Mineral	Durum semolina ($\mu g\ g^{-1}$ dry wt.)	Hard red spring flour ($\mu g\ g^{-1}$ dry wt.)
K	1976	1159
Mg	690	342
Ca	190	200
P	1850	1273

[a] Taken in part from Pomeranz, ref. 18.

Table 12.5 Effect of polar lipids associated with starch on spaghetti cooking quality

Sample	Cooking quality[26]
Untreated semolina	30.0
Defatted semolina	26.0
Defatted semolina reconstituted	23.6
N-propanol extracted starch reconstituted	0.3

The role of phospholipids in durum wheat and durum wheat products has not been investigated. The high mineral content in durum gluten may, in part, be due to phospholipids associated with gluten. The type of structure proposed by Grosskreutz[21] might be suggested, a phospholipid bimolecular leaflet structure, which could be a possible configuration that gives rise to a more extensible durum gluten. However, a distinction should be made between extensibility due to a high gliadin : glutenin ratio, as found in those weak durum varieties with band 42 by polyacrylamide gel electrophoresis of gliadin protein,[22] and 'sheetability' of stronger varieties. Results from this laboratory have indicated that very strong durum varieties, even those from the Mediterranean area, are not as strong as good bread wheat, like hard red spring varieties. This may be explained on the basis of a higher phospholipid content.

Phosphorus-containing compounds like phytic acid and lecithin do not improve cooking quality of spaghetti processed from durum semolina or hard red spring farina (R. R. Matsuo, unpublished results). It has been reported that the addition of sodium phosphates to spaghetti caused deterioration in color, breaking strength and cooking quality.[23] Therefore, it would seem that if phosphorus has any effect on spaghetti quality it is the intrinsic phosphorus associated with specific components, and extrinsic phosphorus exerts no effect.

Lipids have been implicated in surface stickiness of cooked spaghetti.[24] Non-polar lipids readily extractable with n-hexane or petroleum ether, if removed, increased surface stickiness and when added to semolina decreased stickiness. Polar lipids are intimately associated with starch and may be extracted by refluxing with hot n-propanol or n-butanol.[25]

Preliminary experiments carried out at the Grain Research Laboratory indicate the importance of polar lipids. Semolina, first defatted with n-hexane to remove 'free' non-polar lipids, was fractionated into three crude fractions: starch, gluten and water solubles. The fractions were freeze dried.

The starch fraction was extracted with hot n-propanol. The alcohol was removed and the starch dried. Samples were reconstituted and processed into spaghetti. The effect of polar lipids attached to starch on spaghetti cooking quality appears to be significant (Table 12.5).

Extraction of polar lipids obviously affects starch granules to the extent that spaghetti strands lack rigidity in the three-dimensional structure. The amount of phosphorus in durum semolina (Table 12.4) is significantly higher than that in hard red spring flour and the phytic acid level is very much lower. Thus, it would be reasonable to suggest that a much larger proportion of phosphorus in durum occurs in phospholipids. This may be a factor contributing to the differences in cooking quality of spaghetti made from the two wheat classes.

12.6 Stickiness and cooking loss of spaghetti

The amount of material lost during spaghetti cooking is influenced by water hardness[27] as well as by the pH of the cooking water.[28] A recent study by Malcolmson and Matsuo[29] reported the effect of water hardness on cooking loss and surface stickiness. The magnitude of the effect was influenced by the raw material, i.e. durum semolina versus common wheat farina, used in the spaghetti. Stickiness scores were higher with harder waters, and spaghetti processed from red spring wheat was stickier than that made from durum semolina.

The stickiness phenomenon might be related to the level of insoluble amylose–lipid complexes on the surface of starch granules which influence the retrogradation of starch indirectly, perhaps by altering the distribution of water between starch granules and the surrounding continuous medium. If there is a greater concentration of phospholipids in starches from durum semolina than in those from common wheat flour then less retrogradation can be expected in durum semolina.

The amount of material lost during cooking was similar for both types of spaghetti (i.e. spaghetti made from durum semolina and common wheat farina) for various formulations of cooking water with the exception of the very hard well water.[29] Cooking loss of durum spaghetti was not affected by the hard well water but the cooking loss of the common wheat was very high. These results are unlike those of Dexter *et al.*[27] who found high cooking losses in both common wheat and durum spaghetti with very hard well water. The difference in the two studies was the pH of the well water, 8.0 in the earlier study and 7.6 in the later work. This difference may be sufficient to cause the differences noted.

However, the high cooking loss of common wheat spaghetti cannot be readily explained. The higher levels of minerals in durum semolina (Table 12.4) would suggest that amylose, the principal component in the cooking water leached out during cooking,[30] is complexed and remains in the spaghetti. In well water the concentration of the minerals is so high relative to other prepared water, it is difficult to determine which one or

Table 12.6 Mineral composition of cooking waters[29]

Water	Mineral composition (ppm)				
	Ca	Mg	Na	K	SO$_4$
Tap	24.20	7.46	2.47	–	3.40
F1	28.96	3.11	82.41	11.64	14.32
F2	14.48	1.55	41.20	5.82	7.16
F3	32.58	3.59	93.14	13.49	16.55
F4	32.58	3.59	72.60	13.49	16.55
Well	93.00	68.00	153.00	–	170.00

combinations of ions causes the large cooking loss. The composition of the cooking waters used in the study[29] is given in Table 12.6.

A recent study[31] correlated cooking loss with absorption of iodine–amylose complex formed in an aliquot of clarified cooking water. It was found that cooking loss increases with cooking time but the rate of spaghetti breakdown with extended cooking time is greater in common wheat pasta than in durum pasta. The cooking water used in this study was prepared to contain fixed amounts of Ca, Mg, Na, K and SO$_4$ at pH 7.5 and hardness of 96.1. Overcooking tests were not conducted with harder or softer cooking water. The fact that very hard well water caused excessive spaghetti breakdown in common wheat spaghetti would suggest that the rate of breakdown would be greater in harder water. The specific element or elements that may be responsible for the differences in cooking quality between the two classes are not known.

Another element that may have a role in cooking quality is copper. The average copper content of five wheat classes representing over 400 samples of 231 varieties was reported to be 0.49 mg/100 g of wheat.[18] Copper deficiency has an effect on baking quality and physical dough properties.[32] The effect of copper on pasta quality has not been studied but copper has been associated with a protein that causes brownness in pasta.[33] Most of the durum varieties with brownish discoloration tend to have strong gluten with good cooking quality, e.g. many of the durum varieties grown in the Mediterranean region are brownish but with good cooking quality. Thus, if copper improves baking quality, it may also be possible that copper has a role in the textural characteristics of cooked pasta.

The cooking quality of pasta is, like baking quality, governed by many factors. The role of minerals has not been fully investigated; it is suggested that further studies on metal association with lipids, gluten proteins and starch may shed more light on pasta cooking quality.

References

1. International Wheat Council, (1991) *Market Report PMR 193*, London.
2. Pandy, H. N. (1984) Development of multilines in durum wheat cultivars. *Rachis*, **3(1)**, 9–10.
3. International Wheat Council (1987) *World Wheat Statistics*, International Wheat Council, London.
4. Alsberg, C. L. (1939) Durum wheats and their utilization. *Wheat Studies of the Food Research Institute*, Vol. XV, No. 7.
5. Srivastava, J. R. (1984) Durum wheat – its world status and potential in the Middle East and North Africa. *Rachis*, **3(1)**, 1–8.
6. Williams, P. C. (1985) Survey of wheat flours used in the Near East. *Rachis*, **4(1)**, 17–20.
7. Nikolopoulou-Pattakou, V. (1981) Uses of hard wheat and semolina other than for pastas. International colloqium. *Hard Wheat Production and Uses in Mediterranean and European Countries*, Paris, pp. 18–19.
8. Williams, P. C., Srivastava, J. P., Nachit, M. and El-Haramein, F. J. (1984) Durum wheat quality evaluation at ICARDA. *Rachis*, **3(2)**, 30–33.
9. Axford, D. W. E., McDermott, E. E. and Redman, D. G. (1979) Note on the sodium dodecyl sulfate test of breadmaking quality: Comparison with Pelshenke and Zeleny tests. *Cereal Chem.*, **56**, 582–584.
10. Dexter, J. E., Matsuo, R. R., Preston, K. R. and Kilborn, R. H. (1981) Comparison of gluten strength, mixing properties baking quality and spaghetti quality of some Canadian durum and common wheats. *Can. Inst. Food Sci. Technol. J.*, **4**, 108–111.
11. Tipples, K. H. and Kilborn, K. H. (1974) 'Baking Strength Index' and the relation of protein content to loaf volume. *Can. J. Plant Sci.*, **54**, 231–234.
12. Dick, J. W. and Matsuo, R. R. (1988) Durum wheat and pasta products. In *Wheat: Chemistry and Technology*, Vol. II, American Association of Cereal Chemists, St. Paul, Minnesota, pp. 507–547.
13. Sheu, R. Y., Medcalf, D. G., Gilles, K. A. and Sibbit, L. D. (1967) Effect of biochemical constituents on macaroni quality. I. Difference between hard red spring and durum wheats. (1967) *J. Sci. Food Agric.*, **18**, 237.
14. Wyland, A. R. and D'Appolonia, B. L. (1982) Influence of drying temperature and farina blending on spaghetti quality. *Cereal Chem.*, **59**, 199–201.
15. Kim, A. I., Seib, P. A., Posner, E., Deyoe, C. W. and Yang, H. C. (1986) Milling hard red winter wheat to farina: comparison of cooking quality and color of farina and semolina spaghetti. *Cereal Foods World*, **31**, 810.
16. Malcolmson, L. J., Matsuo, R. R. and Balshaw, R. (1993) Optimization of spaghetti quality using response surface methodology: effects of drying temperature and blending hard red spring farina with durum semolina. *Cereal Chem.*, **70**, 417–423.
17. Irvine, G. N. (1979) Durum wheat quality: comments on the International Collaborative Study. Comptes rendu du *Symposium International sur: Matières Premièrs des Pâtes et Alimentaires*, (ed. G. Gabriani and C. Lintus) Conseil National des Recherches D'Italie, Rome, pp. 31–40.
18. Pomeranz, Y. (1988) Chemical composition of kernel structures. *Wheat Chemistry and Technology*, Vol. I, (ed. Y. Pomeranz) American Association of Cereal Chemists, St. Paul, Minnesota, pp. 97–158.
19. Taberkhia, M. M. and Donnelly, B. J. (1982) Phytic acid in durum wheat and its milled products. *Cereal Chem.*, **59**, 105–107.
20. Mecham, D. K. and Mohammad, A. (1955) Extraction of lipids from wheat products. *Cereal Chem.*, **32**, 405–415.
21. Grosskreutz, J. C. (1961) A lipoprotein model of wheat gluten structure. *Cereal Chem.*, **38**, 336–349.
22. Bushuk, W. and Zillman, R. R. (1978) Wheat cultivar identification by gliadin electrophoregrams. I. Apparatus, methods and nomenclature. *Can. J. Plant Sci.*, **58**, 505–515.
23. Evans, G. C., de Man, J. M. and Rasper, V. (1975) Effect of polyphosphate addition in spaghetti. *Can. Inst. Food Sci. Technol. J.*, **8(2)**, 102–108.
24. Matsuo, R. R., Dexter, J. E., Boudreau, A. and Daun, J. K. (1986) The role of lipids in determining spaghetti cooking quality. *Cereal Chem.*, **63**, 484–489.
25. Soulaka, A. B. and Morrison, W. R. (1985) The amylose and lipid contents, dimensions,

and gelatinization characteristics of some wheat starches and thin A- and B-granule fractions. *J. Sci. Food Agric.*, **36**, 709–718.

26. Dexter, J. E. and Matsuo, R. R. (1977) Influence of protein content on some durum wheat quality parameters. *Can. J. Plant Sci.*, **57**, 717–727.

27. Dexter, J. E., Matsuo, R. R. and Morgan, B. C. (1983) Spaghetti stickiness: some factors influencing stickiness and relationship to other cooking quality characteristics. *J. Food Sci.*, **48**, 1545–1551, 1559.

28. Alary, R., Abecassis, J., Kobrehel, K. and Feillet, P. (1979) Influence de l'eau de cuisson, et notamment de son pH, sur les caractéristiques des pâtes alimentaire cuite. *Bull ENSMIC*, **293**, 255–262.

29. Malcolmson, L. J. and Matsuo, R. R. (1993) Effects of cooking water composition on stickiness and cooking loss of spaghetti. *Cereal Chem.*, **70**, 272.

30. Dexter, J. E., Matsuo, R. R. and MacGregor, A. W. (1985) Relationship of instrumental assessment of spaghetti cooking quality to the type and amount of material reused from cooked spaghetti. *J. Cereal Sci.*, **3**, 39–53.

31. Matsuo, R. R., Malcolmson, L. J., Edwards, N. M. and Dexter, J. E. (1992) A colorimetric method for estimating spaghetti cooking losses. *Cereal Chem.*, **69**, 24–29.

32. Flynn, A. G., Panozzo, J. F. and Gardner, W. K. (1987) The effect of copper deficiency on the baking quality and dough properties of wheat flour. *J. Cereal Sci.*, **6**, 91–98.

33. Matsuo, R. R. and Irvine, G. N. (1967) Macaroni brownness. *Cereal Chem.*, **44**, 78–85.

13 Wheat utilization for the production of starch, gluten and extruded products

F. MEUSER

13.1 Introduction and overview

Five years ago in San Diego, USA, a Wheat Industry Utilization Conference took place under the direction of Y. Pomeranz who published the presented papers under the title *Wheat is Unique*.[1] The information gathered in that publication still represents today the present state of knowledge on the structure, composition, processing, end-use properties and products of wheat. This is also true of the topics discussed here, the extraction of starch and gluten from wheat and extrusion cooking, on which the present author contributed two conference papers.[2,3] Therefore, as only a few new findings can be reported, this chapter will concentrate on presenting the current state of knowledge on these topics with special reference to the functional characteristics of wheat gluten and starch. In so doing, the process technological possibilities for the extraction of starch and gluten from wheat flours and wheat wholemeal flours as well as the application of wheat starch and wheat flours for the manufacture of extrusion cooking products will be emphasized. Special emphasis is also given to the gluten-forming properties of wheat proteins and the two starch fractions present in the endosperm which differ not only in size but also in physical and chemical properties.

Using laboratory scale centrifugal separations of doughs, the basic principles for separating starch granules out of dough slurries are explained. It is also shown that the viscous properties of the smaller starch granules differ considerably from those of the larger ones. This difference is of great interest in the industrial extraction of starch, since the yield ratio between A- and B-starches can be regulated by a proper adjustment of the processing steps, whereby the viscous properties of the A-starch are mainly influenced. The difference in the gelatinization properties between large and small starch granules is also evident from the various applications of wheat flours, although in most cases their enzymic activities have a dominant effect on the viscous behaviour of the flours.

On the basis of these results, the possibility of utilizing a wholemeal wheat flour for the extraction of starch and gluten is described using a new process technology developed on a laboratory scale. It is shown that, by extraction of the soluble substances from the wholemeal wheat flour by suspension in water before the separation of the solids, the process water can be adjusted to a dry matter content of approximately 6%. Thus a precondition is given for the economic recovery of these soluble substances and a dry matter yield is obtained which is similar to that obtained in the corn starch extraction process.

In the last section of this chapter, the development of a systems analysis approach to examine the extrusion cooking process is described. The advantages of this approach are presented using the extrusion cooking of soft wheat flours as an example. It is shown that the extrusion cooking process is influenced by the starch content of the ground product. In comparison, other raw material components have only a slight influence on the extrusion cooking process and the resultant extrudate properties. For this reason, certain end-product properties can be better controlled via a suitable choice of extrusion parameters rather than via the selection of ground products from soft wheats which have different quality character-istics.

Since results already published are used here, the various processing and experimental methods will only be described as is necessary to understand the findings.

13.2   Physical characteristics of gluten proteins and starch fractions

The two special features of wheat which will be examined are first, the ability of one part of its protein fractions to swell up with water and form, under the influence of mechanical energy, a cohesive viscoelastic mass termed gluten and, secondly, the characteristic presence of two fractions of starch granules which differ considerably from one another in size and shape. These features will be examined as far as is necessary to understand the application of processing techniques.

13.2.1   Gluten formation from wheat proteins

The formation of gluten from milled products of wheat during dough production is a phenomenon which is still not fully understood, thus making wheat an irreplaceable raw material in the production of a considerable proportion of our foodstuffs. The most important of these are the various baked goods whose manufacture depends on the ability of the gluten to form a film with gas-retaining properties. The film formation is dependent on the production of agglomerates which can be extended into strands and films. The agglomerates result from the hydration of protein molecules and the building of a network of inter- and intra-molecular bonds.

This type of gluten formation and elasticity has recently been made visible by Amend and Belitz[4,5] using various microscopic techniques (light,

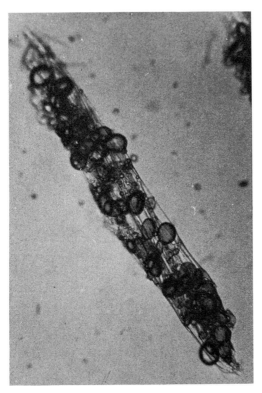

Figure 13.1 Light microscopic photograph of an extended particle from wheat flour; gluten visible in the form of strands, extended between the starch granules in the direction of stress.

transmission and scanning microscopy). Although, the interpretation of the photographs must be treated with some caution due to the presence of possible artifacts formed during the preparation of the material, the author has been able to develop, on the basis of these pictures, a clear model of gluten's formation and elasticity.

According to this model, the native gluten proteins, present as irregular globular structures, build a three-dimensional network when the flour is mixed with water. The individual strands of this network are extended during kneading and simultaneously stretched on account of their elasticity (Figure 13.1). As a result of the tensile stress, small plate-like films (Figure 13.2) are formed at the junctions of the protein strands. These build up in layers to form a type of membrane. Strong stretching tears the films or membranes (Figure 13.3) but the layers remain intact.

This plausible model is also of great importance in the explanation of modern processes of gluten agglomeration and starch extraction out of the dispersions. This will be discussed in more detail with the presentation of the

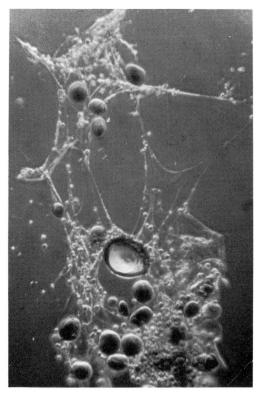

Figure 13.2 Light microscopic photograph of a bidirectionally extended particle from wheat flour; strands of gluten strongly extended and partly torn apart into films at branching points.

present day technology. At this stage, the basic groundwork will be outlined with the help of a few simple experiments.

13.2.2 Centrifugal separation of a wheat flour dough

In wheat flour doughs, the starch granules are embedded in a viscoelastic matrix which is principally formed from the hydrated and agglomerated gluten proteins together with hydrated pentosans. For the extraction of starch from wheat flour doughs, it is of especial interest to realize that there is a difference in density between the starch granules and the matrix. This fact can be utilized partially to separate the dough into its component masses by means of centrifugal forces.[6]

When such a separation is carried out in a centrifuge beaker, two to four clearly separable layers are formed, depending on the chosen proportions of water and flour (W/F ratio). These can be separated from one another by simple decantation and/or syphoning off. Figure 13.4 shows the result of

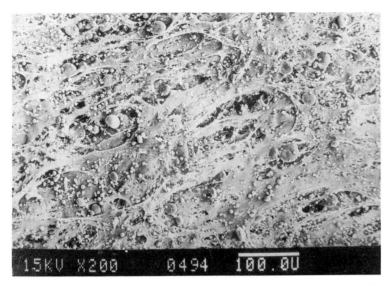

Figure 13.3 SEM photograph of a fully developed dough; gluten films oriented in the direction of stress.

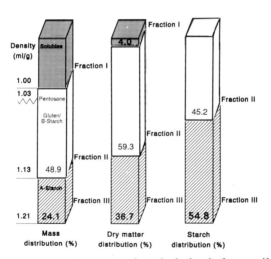

Figure 13.4 Distribution of mass, dry matter and starch of a dough after centrifugation. Dough dry matter, 38%; centrifugal acceleration, 3000 g.

such an experiment for which water and flour (ash content 0.6% dry mass (dm)) were mixed to form a smooth elastic dough with a solids content of 38%. This dough was centrifuged at 3000 g.

Four layers of differing density were obtained. *In situ*, the upper layer, which consisted mainly of water which was not bound by the dough but which contained soluble compounds from the flour, had a density of $1.0\,g\,cm^{-3}$. The lower layers had increasing apparent densities of 1.03, 1.13 and $1.21\,g\,cm^{-3}$, respectively from top to bottom on account of their contents of insoluble substances especially starch granules and fibres. The difference in apparent densities between the third and fourth layers is of significance. Taking into account that starch-free wet gluten and starch granules have densities of approximately 1.1 and $1.4\,g\,cm^{-3}$, respectively, the basis is given to separate starch granules out of dough.

The second and third layers, which consisted essentially of gluten, the insoluble pentosans and the non-separable starch granules, were mixed together before the distributions of mass, dry mass and starch granules were determined so that only three fractions had to be referred to in the results. The percentages of the total mass in fractions I, II and III were 27.0, 48.9 and 24.1, respectively which means that, when taking account of the densities, the total volumes were 29.8, 48.2 and 22.0%, respectively. The dry matter was present to 4% in fraction I and to 59.3 and 36.7% in fractions II and III, respectively. Since the dry matter in fraction III consisted predominantly of starch, this represented 54.8% of the starch present in the flour. Thus, in this experiment, only a little more than half of the starch could be separated from the matrix despite the difference in densities.

This led to questions regarding the operating mechanism of the separation. From further results obtained from the test series belonging to this experiment, it could be seen that the starch cannot always be separated from wheat doughs by means of centrifugal forces since the separation is also dependent on the viscoelastic properties of the matrix. In this respect it was found that the viscosity and elasticity of the doughs rose rapidly with smaller W/F ratios. With a W/F ratio of 0.7 (approximately 50% d.m. in the dough), the resistence, due to the viscosity and elasticity of the matrix, was so large that even with a centrifugal force of $3000\,g$ no starch could be separated from the dough.

A gradual reduction in the apparent density of the dough by increasing the W/F ratio eventually gave a value at which the separation led to a maximum enrichment of the dry matter in fraction III (Figure 13.5).[6] By further reductions in the apparent density the starch enrichment in fraction III was again lower although, at the same time, the mass proportion of fraction II remained practically constant up to a W/F ratio of about 1.3. In the region of this W/F ratio the distribution between fractions II and III approached a limiting value asymptotically. This was characterized by the fact that approximately 50% of the available starch was to be found in fraction III.

This fact has led to the question, why, despite the relatively large differences in apparent densities between fractions II and III in comparison to tests with a smaller W/F ratio, the decline in the separation efficiency, which is characterized by the dry matter weight of starch in fraction III, was associated with the smallest value? Since a relatively high proportion of the

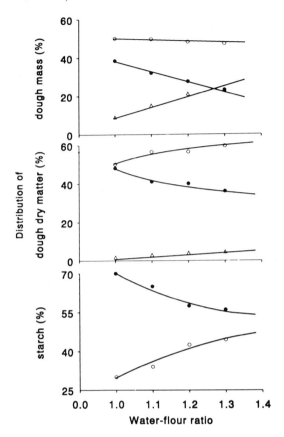

Figure 13.5 Distribution of mass, dry matter and starch of doughs prepared under different water–flour ratios after centrifugation. Fraction I, △; fraction II, ○; fraction III, ●.

starch mass was always to be found in fraction II, it was postulated that the separation of the starch from the matrix in the dough is also dependent on the volumetric ratio between the matrix and the starch. The assumption was based on the consideration that, during centrifugation, the pressure applied to the matrix per unit surface area must be dependent on the volumetric concentration of the starch mass in the matrix mass.

In the example shown in Figure 13.5, the volume ratio of starch to matrix, after deduction of the water not bound to the starch, was 1.33. During centrifugation this value fell to 0.6 because of the separation of the starch out of the matrix. As a result, the pressure applied by the starch mass on the matrix was only half as much as it was at the start.

If the gluten is now considered to be a network composed of layers of thin films, as described by Amend,[4] it is understandable that this can only be penetrated by the starch granules when the applied pressure is greater than the resistance resulting from the viscoelastic properties of the matrix. It follows that the larger starch granules must be able to pass through the network more easily than the smaller ones since they apply a greater force per unit area than do the smaller granules. In theory, whilst the cross-section of the granules increases as the square of their radius, the mass increases by a power of three. For this reason the size distribution of the granules in the fractions should differ from that in the dough. Thus, in the described experiment, there should be a greater proportion of large granules in fraction III than in fraction II.

The evidence for this assumption was obtained from size distribution curves for portions of fractions II and III. These experiments were carried out, similarly to those previously described, but with flours from varietally pure soft wheats which had an ash content of approximately 0.6% (extraction rate approximately 78%). After the first centrifugation of the doughs formed from the flours, fraction I was discarded and fractions II and III were treated further.

The starch in fraction III was purified, as for normal commercial starches, by repeated washing through a sieve (pore size 63 μm) followed by centrifugation of the filtrate. The starch so-obtained was labelled fraction A. Fraction II was recentrifuged to extract a further portion of starch which was then purified, as described before, and labelled fraction B. The starch still remaining in fraction II was then also washed out, purified and labelled fraction C.

In order to be able to balance the starch extracted, the starch content of fraction C was determined. For the same reason the dry matter contents of the wet masses were measured. The wet masses were then suspended in water in order to measure the size distribution of the starch granules they contained using the Coulter-Counter measuring technique. The results can be seen in Figure 13.6.[7]

It is immediately apparent that the three distribution curves for fractions A, B and C are different. They differ, above all, in their rise and shape. Both criteria demonstrate that, on examination of the extracted starch, which is about 85% of the total, the mass fraction of the granules with a particle size of about 5 μm accounted for about one third of the total mass of granules of size up to 10 μm. The starch within the granule size range of 0 to 10 μm was 19.6% of the total examined. Since the non-extracted starch granules, which accounted for 15% of the total, were mainly to be found in the sieved-out pentosan fraction, it was assumed that their size distribution was similar to that of mass C. Thus the figure for the proportion of starch granules less than 10 μm could be adjusted to approximately 25%. It should be mentioned that practically the same result was also obtained for all other soft wheat flours examined.

To summarize the results of these experiments, it can be said that the resistance due to the viscoelastic properties of the matrix effects the

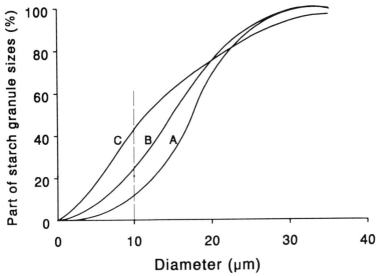

Figure 13.6 Size distribution of starch granules in fractions extracted from a German soft wheat variety. Part of granule sizes < 10 μm in fractions A–C: 19.6%. Part of the total starch of the flour: (A) 53.7(%); (B) 11.4; (C) 19.9.

separation of the starch granules from the dough quantitatively and qualitatively. The qualitative effect stems from the selective distribution of the smaller starch granules within the fractions. However, the question remains unanswered as to why the separation effect was worse when the W/F ratio was lower.

A cause for this contradictory result was sought in possible different agglomerations of the gluten proteins. This was based on the consideration that the elasticity of the gluten in the doughs could be different due to the intensity with which it was kneaded depending on the W/F ratio. It is known that a dough with a smaller W/F ratio can be kneaded more intensively than one with a larger ratio. Under the influence of the kneading, the dough, and thus also the gluten, develops its elastic properties to a maximum value.

If this concept is transferred to the previous experiment, it implies that, with increasing W/F ratio in the doughs, the gluten had become gradually more elastic since the doughs were produced under otherwise identical conditions. However, this assumption could not easily be tested since the gluten would have had to be extracted from the soft doughs with the same degree of elasticity that it had attained during mixing. Therefore, in order to overcome this problem, a simple comparative trial was made.

For this experiment, two doughs, (a) and (b), were produced. Their W/F ratios where 1.0 and 1.2, respectively, and their dry mass fractions were 45% and 38%. A portion of dough (a) was thinned with water in the same mixer

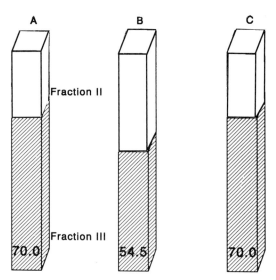

Figure 13.7 Extraction of starch from differently prepared doughs by centrifugation at 3000 g. Dough preparation: water–flour ratio and d.m. (%): (A) 1.0(45); (B) 1.2(38); (C) 1.2(38) (diluted dough (A)).

until the W/F ratio was the same as for dough (b). The three doughs were then centrifuged simultaneously at 3000 g.

As expected, there was a clear difference in the degrees of separation for doughs (a) and (b) (Figure 13.7). Surprisingly however, the thinned out dough (c) behaved similarly to dough (a). Since the results could only be due to differences in the gluten elasticity between doughs (a) and (b) and these, in turn, stem from differences in the kneading intensities during production due to the different W/F ratios, the conclusion is that the elasticity of the gluten has a decisive effect on the separation efficiency of the centrifugation.

One explanation for this influence can be deduced from the results of Amend.[4] It must be assumed that, during passage through the matrix, the starch granules stretch the membranes so far that, at the limits of their elasticity, they tear open or become detached. After a granule has passed through, the membranes contract together again elastically and build a new continuous structure. This differs from the previous one in that it is shrunken by a volume equivalent to that of the separated starch granule. The separation effect is thus dependent on the elasticity of the gluten and it obviously increases with increasing gluten elasticity. The elasticity only needs to be large enough that the gluten does not unravel during stretching. The separation efficiency is supported by the portion of free water, whose function could be described as that of a lubricant. The separation continues until the mass of the starch granules is no longer sufficient to be able to stretch the membranes to breaking point.

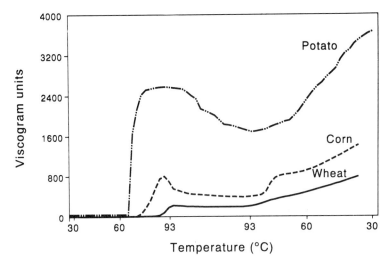

Figure 13.8 Brabender viscogram curves of various starches. Starch concentration: 7%; pH = 7.0.

This description of the separation efficiency can be fully transferred to the continuous extraction of starch from dough suspensions, which will be discussed in more detail later. Beforehand, the different properties of the starch fractions should be examined in order to establish a connection to the second topic of this chapter which is concerned with the extrusion cooking of wheat flours and wheat starches.

13.2.3   Physical characteristics of wheat starches

The most important physical characteristic of all starches is their ability to swell up and form a gel when warmed with water. Different starches vary considerably with regard to the expression of this characteristic. The differences can be presented, for example, by the change in viscosity under constant monitoring conditions (Figure 13.8). Thus, at the same concentration, wheat starch is far less viscous than potato or maize starch at its viscosity maximum. It also begins to gelatinize at a higher temperature than the others; the viscosity maximum is also attained at a higher temperature. Similar differences are observed for the enthalpy of gelatinization using the differential scanning calorimetry (DSC) thermogram (Figure 13.9). In this respect it is especially worth mentioning that starches display further enthalpy peaks above 100°C. These peaks can be attributed to amylose lipid complexes of the starches and dense structural arrangement of their molecules.

The course of the gelatinization of wheat starch, shown here with viscograms and thermograms, is otherwise only applicable to the so-called

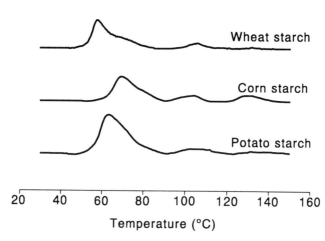

Figure 13.9 DSC thermograms of wheat (ΔH:12.2 J/g), corn (11.3) and potato (20.1) starch.

primary starches as obtained from wheat flour on an industrial scale. This fraction consists mainly of the larger starch granules in the wheat. The fraction with smaller granules accumulates with far less purity in the so-called secondary starch. This is usually dried on drum dryers together with mucins to form pregelatinized starch.

Of interest is the fact that the viscous behaviour of the smaller starch granules differs considerably from that of the larger ones. This difference can be partly traced back to the composition of the starch granules. In consideration of the previous work and in order to demonstrate these associations, three different starch fractions were obtained from a wheat flour by centrifugal separation and rinsing. Scanning electron micrograph (SEM) photographs, as well as viscograms and DSC thermograms, were taken of these starches. Added to this, the starch fractions were characterized analytically according to their starch, protein, lipid and ash contents.

To obtain the starch fraction, a wheat flour (ash content 0.6% d.m., extraction rate 78%) was premixed with water to form a dough (W/F ratio 1.0, approximately 45% d.m.). As already described, the dough was centrifuged at 3000 g. The fractions obtained in the centrifuge beaker were separated from one another and fraction I was discarded. The first starch fraction (SF I) was obtained from fraction III by further rinsings and centrifugations of the suspension. Fraction II was washed thoroughly over a sieve (pore size 63 μm). A slimy mass remained behind on the sieve. This consisted of insoluble pentosans in which a large portion of the smaller starch granules was embedded. Those larger starch granules which were not separated out of the matrix of gluten and pentosans by centrifugal force, ended up in the filtrate together with some of the smaller starch granules. This starch mass was further purified by repeated suspension in water

Table 13.1 Analytical characterization of the extracted starch fractions

Content	Starch fraction		
(% dm)	SF I	SF II	SF III
Starch[a]	99.5	93.0	88.9
Protein	0.7	0.7	1.2
Fat[b]	1.0	1.0	1.4
Ash	0.18	0.21	0.28

[a] Enzymic method: ICC Standard No. 128. [b] Acid hydrolysis method: AOAC 14.019.

followed by centrifugation. After the last suspension, three stable layers formed in the centrifuge beaker and were then separated from one another. The upper layer consisted of a further part of slimy mass with embedded small starch granules. Below this was a fraction which consisted principally of small starch granules. It formed the third starch fraction (SF III) and the lowest layer formed the second starch fraction (SF II). The starch fractions contained approximately 84% of the starch present in the flour. The approximate distribution amongst the individual fractions, SF I, II and III was 61, 19 and 4%, respectively.

The analytical composition of the starch fractions is given in Table 13.1. The relatively large lipid content of the small starch granules is worth mentioning since this has a considerable influence on the viscous behaviour of the small starch granules.

The selective action of the centrifugal separation is clearly recognizable in the SEM photographs of the starch fractions (Figure 13.10). Even though larger starch granules were also present in the SF III, the main mass of the starch in this fraction consisted of small starch granules. For this reason, their gelatinization characteristics are to be seen as representative for this fraction.

The viscogram and the DSC thermogram confirm the large differences in the gelatinization characteristics which exist between large and small starch

Figure 13.10 SEM photographs of wheat starch fractions.

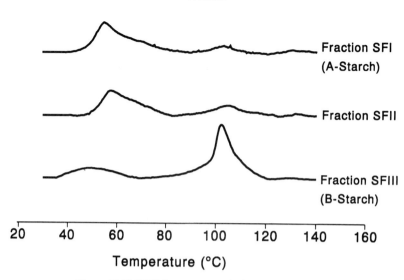

Fraction SFI
(A-Starch)

Fraction SFII

Fraction SFIII
(B-Starch)

20　40　60　80　100　120　140　160

Temperature (°C)

Figure 13.11 DSC thermograms of wheat starch fractions.

granules (Figure 13.11). It is therefore understandable that the application properties of A-starch are also dependent on the ratio of large to small starch granules. This ratio can be controlled to a certain extent by process technology during wheat starch extraction.

The differences between the gelatinization characteristics of large and small starch granules are also important in the applications of wheat flours.[8] Besides the ratio of large to small granules, which is partly genetically determined and partly apparently dependent on the growing conditions, the enzymic activity of the flours also plays an important role. For example, two flours, with the same ash content and extraction rate, but varied α-amylase activities will give rise to different degradation rates for the starch molecules during gelatinization. These differences are evident in the amylograms by variably large viscosity maxima, which are also attained at different temperatures.

The influence of the enzyme activity on the degree of gelatinization of the starch can be simply demonstrated by inhibiting the enzymes with mercuric chloride ($HgCl_2$) solution. To illustrate this, Figure 13.12 shows the amylogram curve for a wheat flour (ash content 0.6%) and a rye flour (ash content 1.0%). The considerable differences make clear, above all, the influence which the climatic conditions can have on the gelatinization characteristics of a raw material. In this respect, the differences in the viscosity maxima, as well as in the gelatinization temperatures for the maxima, can be included as reference values for the estimation of the enzymic influence.

It is well known that flours with an increased enzyme activity are neither

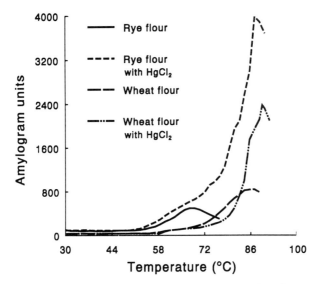

Figure 13.12 Brabender amylogram curves of wheat and rye flours.

suitable for baking nor for starch production. On top of which, in cereals with sprout damage, the starch has already suffered some degradation of the stalk. For this reason, the flours from such raw materials are also less suitable for extrusion cooking purposes, even though the enzymes would be inactivated within a few seconds, and thus are scarcely able to degrade the starch, as they pass through a temperature gradient from 20°C to well over 100°C in the extruder barrel. This will be dealt with again later.

13.3 Starch production from wheat flours

The possibility, already mentioned, of separating wheat starch from doughs centrifugally has been known for a few years. The separation occurs continuously by use of decanting centrifuges.[2] These are fed a dough in the form of a thinned out dispersion in water.

The main advantage of centrifugal separation, as opposed to washing, is that the main fraction, the A-starch, is separated out of the diluted dispersion in a matter of seconds, whereas it takes up to half an hour with the washing process. This shortening of the retention time of the mass in the separation system in combination with the small solids and water volume in the system per unit of time (approximately 20 times smaller than in the Martin process) is extremely advantageous with regard to the process water flow and its pH.

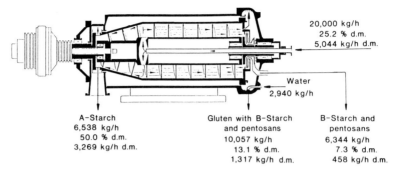

20,000 kg/h
25.2 % d.m.
5,044 kg/h d.m.

Water
2,940 kg/h

A-Starch	Gluten with B-Starch	B-Starch and
6,538 kg/h	and pentosans	pentosans
50.0 % d.m.	10,057 kg/h	6,344 kg/h
3,269 kg/h d.m.	13.1 % d.m.	7.3 % d.m.
	1,317 kg/h d.m.	458 kg/h d.m.

Figure 13.13 Centrifugal decantation of a flour–water dispersion into three phases. Reproduced from Witt, W. (1988).

Recent advances in the field of centrifugal separation technology now make it possible to separate the suitably diluted dispersion into three solids-containing fractions. This is especially advantageous with regard to the necessary refining of the gluten, since in this way solid and dissolved substances are removed from the fraction containing the gluten. Splitting a dispersion with a decanter operating according to this principle is illustrated in Figure 13.13 (W. Witt, 1988, personal communication).

In this context, it should be mentioned that the process water, after separation of the B-starch and pentosans, contains most of the soluble flour constituents. Process water is used in the factory which uses the process according to the mass balance given in Figure 13.13 partly to dilute the dispersion before decanting. By this means, a constant concentration of dissolved substances of 3.0–3.3% can be adjusted in the process water at a constant process water volume. The product specifications do not permit a higher concentration so that the process water volume to be treated as effluent is precisely defined.

An appreciable increase in the process water concentration, in order to reduce the effluent volume, would require that the flour dry substance be subjected to a completely different preparation which would alter the installation concept accordingly. A logical approach would be to extract the soluble substances before agglomeration of the gluten. This technique seems to be realizable since, with homogenizers, the gluten can be agglomerated in flour–water suspensions obtained by concentrating highly diluted flour–water suspensions. This has been shown in laboratory-scale experiments.

For this purpose we built a laboratory plant in which it is possible to process 1.8 kg of wholemeal wheat flour per hour. Wholemeal flour, which was obtained by grinding a suitably conditioned wheat using an Ultra-Rotor,[9] was used in order to have the whole endosperm available for the starch extraction. This is not the case when white (processed) flour is used

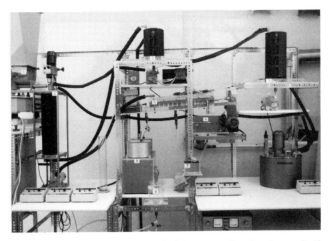

Figure 13.14 Photograph of the main parts of a laboratory plant for the production of starch and by-products from wholemeal wheat flour.

since a portion of the endosperm always ends up in the bran during dry milling. In principle, of course, it is also possible to use white flour for the newly developed process. However, since additional costs are involved in the process, especially in the centrifugal separation of the extracted endosperm slurry and the recovery of dissolved substances, these must be compensated for by improving the efficiency of the total process, starting with the milling of the wheat.

Figure 13.14 shows the main parts of the laboratory plant for processing a wholemeal flour product. Clearly visible are the container, in which the soluble substances are extracted from the wholemeal flour by suspending in water, and the sieve on which the fibres are separated out of the suspension. Further details are provided in Figure 13.15 which shows a simplified flow diagram for the newly developed process. The laboratory plant process differs from that shown in this flow diagram in that the gluten agglomeration does not occur continuously as is intended for the industrial process. Instead the decanted endosperm slurry is made into a paste in a laboratory mixer. This is necessary since there has been until now no suitable machine capable of continuously processing a mass of about 4 kg h^{-1} in this system.

However, the transfer of the laboratory process to the industrial scale is not limited because of this since it has already been demonstrated that the gluten can also be agglomerated from a suspension of suitable solids. An agglomeration sufficient to achieve a separation, which is characterized by a certain elasticity of the gluten, can be attained, for example, with a solids concentration of 38%. The agglomeration can be undertaken in similar systems to those presently used in modern wheat starch factories.

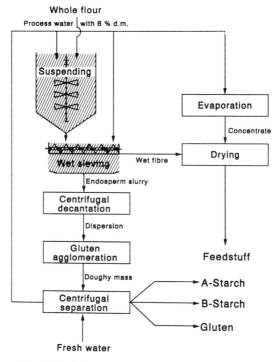

Figure 13.15 Simplified flow diagram for a new wheat starch process.

In these systems, white flour and water are mixed together in the first step, for example with a continuously operating mixer. In this process step, an agglomeration of the gluten has already occurred to an extent. The resultant, easily pumpable mixture has a solids content of approximately 45%. For further agglomeration, the mixture is fed through a high-pressure homogenizer. On exiting the homogenizer, the mass is continuously mixed with water and the solids content adjusted to approximately 25%. It is then fed into the decanter centrifuge.

The necessary gluten elasticity is achieved on passage of the mixture through the pumps and piping and especially at the valve of the homogenizer. Whilst difficult to prove, it is easy to imagine that the energy transfer in the slit of the valve and throughout its length is responsible for the two-dimensional stretching of the gluten proteins which is important for film formation. However, this can be assumed to be the case judging from the increase in the viscoelasticity of the mass between the inlet and outlet of the homogenizer.

In the newly developed process, one must be careful that, by centrifugal separation of the hydrated endosperm particles, the swollen gluten proteins do not become separated from the starch granules because the handling of the mass would otherwise be difficult. The separation must occur in such a way that there is a homogeneous distribution of protein particles and starch granules in the mass which thus remains easily pumpable.

With the laboratory plant, it could be shown that a slurry out of wheat wholemeal flour is sievable when no gluten formation occurs on the sieve. To achieve this, the sieve must have special constructive features. These must be so designed that an energy transfer, in the form of a kneading effect, does not occur. For example, a kneading effect can be caused when the endosperm particles experience a shear stress or are compressed when passing through the sieve. Such effects were observed to occur when a vibrating sieve is used. Such sieves are thus unsuitable for the wet sieving of wholemeal flours.

The sieve used in the laboratory plant consists of a semicircular trough in which a screw conveyor (length 500 mm, diameter 39 mm, pitch 32 mm) rotates. The sieve is divided into two zones. Product feeding and sieving occur in the first zone. In the second zone, the sieve residue is washed through with recirculated water in order to obtain as complete a separation as possible of the remaining endosperm from the fibres. The sieve mesh is made from monofilamentous nylon gauze with a pore size of 250 μm. The underside of the sieve is brushed by oscillating rakes with teeth consisting of Teflon strips. This is to prevent the sieve from blocking-up. This sieve is also capable of functioning in permanent operation.

The special advantage of the processing stage achieved with the laboratory plant is that the concentration of the substances dissolved in the process water is double that attainable in the most modern industrial plant (Table 13.2) (F. Althoff, 1991, personal communication). Thus, in consideration of the costs of effluent treatment and disposal, an economic recovery of the soluble solids is possible whilst at the same time solving a considerable environmental problem. The solids yield from the wholemeal wheat flour can be as high as that attained by the production of corn starch. For the laboratory scale trials a value of between 98.5 and 99.2% was attained.

We are presently investigating the transferability of the process to an industrial scale.

Table 13.2 Yield of dry matter from whole flour during an 18 h test run of the laboratory system compared with results of a modern industrial process run on wheat flour[a]

Extraction	Concentration of process water (%)	Mass distribution (%) Fibre[b] Bran	Endosperm/flour	Dry matter yield (%)
Laboratory system[c]	5.8–6.7	13.6–19.3	82.7–79.4	98.5–99.2
Industrial process[d]	3.0–3.3	20.0–22.0	80.0–78.0	92.0–92.2

[a] Source: F. Althoff (dissertation in preparation). [b] Wet sieving (mesh width 250 μm). [c] Figures belong to the running time from the 6th to the 18th hour. [d] Calculation includes the bran as by-product from wheat milling.

13.4   Extrusion cooking of wheat starches and wheat flours

Extrusion cooking of wheat starches and wheat flours has become a considerably important process technique in the starch and snack food industries. For this, extruders of various design and size are used to produce, for example, pregelatinized starches, starch derivatives, snacks, breakfast cereals and flat breads. The extruders are applied to mix, plasticize, chemically modify and shape raw materials and mixtures of these which, in most cases, contain either large amounts of starch and/or protein. Wheat-based raw materials are usually extruded in mixtures with sugars, proteins, fats and dietary fibres to achieve specific sensory and nutritional final product characteristics.

Under extrusion cooking conditions, the raw materials show a rather complex reaction behaviour which is connected to the transition of the plasticizable components, mainly the starch, to a plast. The plastification of starch by extrusion cooking must be seen in close connection with its gelatinization characteristics. However, whilst in many applications, the gelatinization of starch occurs by relatively slow thermal energy transfer with a lot of water and with few, if any, shear forces, exactly the opposite conditions are present during extrusion cooking. The mass to be extruded is simultaneously warmed and plasticized by frictional forces resulting from the shear stresses experienced during a residence time of only a few seconds in the extruder barrel. The amount of heat energy absorbed, leading to a rise in temperature of the mass, is inversely proportional to its water content. The essential function of the water molecules is to break up the hydrogen bonds existing between the starch molecules by hydration with the help of mechanical and thermal energy. This process is comparable to that occurring during the swelling of starch when it is boiled in an excess of water.

The hydration process leads under extrusion cooking conditions to plastification of the starch which establishes a strong increase in viscosity of the mass. The increase in viscosity creates a resistance force which works against the mechanical energy input. The two opposing forces exist in a reciprocal relationship which, via the release of the mass through the die and the change in its viscous behaviour, produced for example by breakages to the main valence bonds in the starch molecules, can be brought into a dynamic equilibrium which gives rise to certain end-product characteristics.

This description of the extrusion cooking process gave us the idea that the adjustment of the equilibrium system could be mathematically formulated as a functional relationship and thus that the whole process could be described in systems analysis terms. To illustrate this line of thought, results are presented in Table 13.3 from extrusion cooking trials with commercial wheat, corn and potato starches. All trials were carried out under constant conditions using a twin-screw extruder (Werner & Pfleiderer: Type ZSK 25).

Table 13.3 Extrusion cooking of various starches

Experiments and results		Wheat starch	Corn starch	Potato starch
Gelatinization enthalpy	$(J g^{-1})$	12.2	11.3	20.1
Extrusion parameters				
Water content	(%)	18	18	18
Throughput rate	$(kg h^{-1})$	8	8	8
Screw speed	(min^{-1})	174	174	173
Barrel temperature	(°C)	90	90	90
System parameters				
SME	$(W h kg^{-1})$	120	120	213
PT	(°C)	130	128	138
MRT	(s)	32	33	32
Pressure	(bar)	70	55	103
Product parameters				
Specific volume	$(ml g^{-1})$	3.9	5.6	1.7
Solubility	(%)	52	56	85
Cold paste viscosity	(Pa s)	0.627	0.378	0.226

It is clearly noticeable that the extrusion parameters (moisture content of the mass, mass throughput, screw rotation speed, barrel temperature) give rise to definite values for the system parameters (specific mechanical energy input: SME, product temperature: PT, mean residence time: MRT, extrusion pressure: EP) which are then mirrored in the extrudate characteristics (specific volume, solubility and cold paste viscosity).

The large differences which exist between the gelatinization processes of potato, wheat and corn starches, as measured by the gelatinization enthalpies, also effect the plastification of the masses during extrusion cooking. As expected for selected extrusion conditions, the potato starch attains the highest viscosity in the barrel chamber as can be seen from the EP reached under constant throughput. The masses also expand differently depending on their attained viscosities. Finally, the transformation of the starch granules into a plastified mass results in physical properties of the extrudates which differ widely from those of the native starch granules. The extrudates show properties which are similar to those of pregelatinized starches. They swell up in cold water and are partially soluble.

The results of the trials clearly demonstrate that the process of extrusion cooking is describable using systems analysis. Therefore, we developed a systems analysis model (Figure 13.16) which distinguishes between independent process parameters, system parameters and target parameters.[10,11] The principle validity of this model was verified by carrying out extrusion experiments with wheat starch according to statistical experimental designs to describe the functional relationships, first, between process variables and system parameters and, second, between process variables and end-product properties.

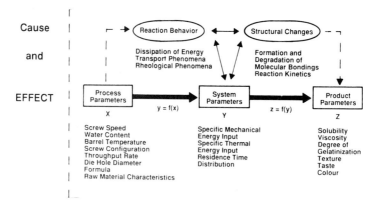

Figure 13.16 Systems analytical model for the extrusion of starch-containing food materials.

The advantage of the mathematical representation of the existing interrelationships between process and system parameters on the one hand and system and target parameters on the other, lies in the fact that the energy input and the end-product properties resulting from the adjustment of an arbitrary combination of process parameters can be calculated. The reverse application of the procedures makes it possible to adjust the process parameters according to any given set of end-product properties or characteristics. Thus the preconditions are fulfilled and enable use of the systems analytical model to develop and optimize products as well as to optimize and scale up extrusion cooking processes.

We recently summarized the results which we have obtained in the last few years whilst developing and using this approach.[12] However, the publication does not deal with the different behaviour of various starches during extrusion cooking. In an initial approach, the integration of these differences in the systems analysis investigation of extrusion cooking can be made by assuming material specific constants. This does not result in any problems provided that the development of the viscosity in the extruder can be related back to that of one particular starch type. It is more difficult, however, when different starches in varied amounts are mixed together as for example in the production of snacks. In such cases, the energy input, dependent on the viscosity formation in the barrel, must be measured specially. This can also be influenced by other components and/or the mixture recipe. The latter influence will be examined here using, as an example, the extrusion cooking of wheat flours of varied extraction rates as well as from wheat flours with various quality characteristics.

In order to carry out the necessary trials, six flours, with different ash contents, were produced in a commercial mill (ash contents: 0.46, 0.57, 0.89, 1.11, 1.59, 1.77% d.m.).[13] The starch content of the flours fell with increasing ash contents (starch contents: 80.9, 79.3, 76.0, 71.4, 67.5,

64.7% d.m.). The extraction rates were in line with the ash contents, lowest for the flour with the smallest ash content. The flour with an ash content of 1.77% d.m. was a wholemeal flour. Further, twelve varietally-pure wheats with varied quality characteristics were ground in an experimental mill (Bühler: Type MLU-202) to produce flours with an ash content of 0.5% d.m. The starch content of these flours was around 76% d.m. The flours were analytically characterized with respect to other quality criteria (e.g. falling number, soluble protein, amylogram with and without added $HgCl_2$).

The flours were extruded, with a twin screw extruder (Werner & Pfleiderer: Type Continua 37), according to a factorial experimental plan. The necessary adjustments to the system parameters as well as the resultant end-product characteristics (e.g. specific volume, expansion index, browning, cold paste viscosity and sediment volume) were measured or analytically determined. With the research results so obtained, the coefficients of the used second order polynomial regression equations[14] were calculated. The results of the calculations were displayed graphically. The regression coefficients were also evaluated with the help of the *t*-test.

Because of the large variation in extraction rate, resulting from the wide range of component composition of the flours, the extraction rates were given as the function of a process parameter on which the energy input was dependent to a certain degree. The extrusion of the flours with the lowest and highest ash contents led to a difference of 2°C in the PT and 14 W h kg^{-1} in the SME (Figure 13.17), whereby the higher values were for the flour with the lowest ash content. The cause of these differences lay mainly in the starch content which increased as the ash content decreased. As a result, the plastified mass in the extruder barrel had the lower viscosity, so that, under otherwise constant conditions, less energy could be transferred due to the lower resistence force.

Under constant energy input conditions, the influence of the extraction rate on the energy input resulted in extrudates with varied characteristics. This is because, under otherwise constant conditions, a certain extraction rate gives rise to a specific energy input. The extrusion of a flour with a divergent higher or lower extraction rate thus results in a divergent mass viscosity for the same energy input. In consequence, the differences in viscosity give rise to various differences in extrudate characteristics. For example, the solubility (Figure 13.18) and the browning of the extrudate increased with the extraction rate, whilst the cold paste viscosity and the sediment volume decreased.

Finally, the extrusion trials with flours from varietally-pure soft wheats showed that the quality characteristics of the raw materials had no influence on the energy input. On the other hand, the extrudate properties were influenced, to a slight extent, by the raw material characteristics. For

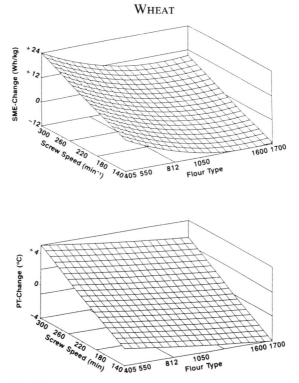

Figure 13.17 Influence of screw speed and flour type on changes in SME and PT.

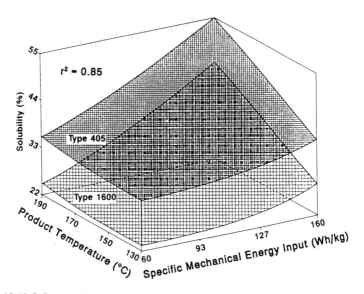

Figure 13.18 Influence of energy input on solubility of extrudates made from different wheat flours.

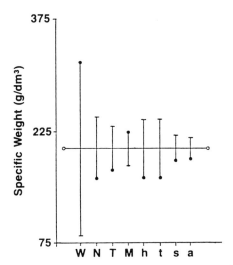

Figure 13.19 Influence of extrusion parameters and raw material characteristics on the specific weight of extrudates made from wheat flours. W = product moisture content (%), T = barrel temperature (°C), M = mass flow (kg h^{-1}), N = screw speed (min^{-1}), h = soluble protein (% dm), t = particle size < 80 μm (%), s = particle size < 100 μm (%), a = starch content (% ds).

example, there was a certain relationship between the content of water-soluble protein in the flours and the specific weight (Figure 13.19) as well as the hardness of the extrudate. Both quality criteria show the larger values for the smaller concentration of soluble protein in the flour and vice-versa. On average, the specific weight was lowered by 40 g dm^{-3} when the amount of soluble protein differed by 0.5% d.m.

By similar trials with rye flours, it was also found that their falling number correlated with the expansion index of the extrudate. Although this was not observed in the extrusion of wheat flours, it is an indication that the enzymic condition of the cereal and its influence on the starch structure can affect the extrusion behaviour of the flour.

In general, these research results show that the influence which raw material qualities have on the extrusion behaviour of flour and on resultant extrudate characteristics, is small in comparison to the effects of mass flow, screw rotation speed, product moisture content and barrel temperature. Therefore, certain end-product characteristics can be better controlled by a suitable choice of these extrusion parameters than via the selection of flours from soft wheats with particular quality criteria.

References

1. Pomeranz, Y. (1989) *Wheat is Unique*, American Association of Cereal Chemists, St. Paul, Minnesota.

2. Meuser, F., Althoff, F. and Huster, H. (1989) Developments in the extraction of starch and gluten from wheat flour and wheat kernels, in *Wheat is Unique*, (ed. Y. Pomeranz) American Association of Cereal Chemists, St. Paul, Minnesota, pp. 479–499.
3. Van Lengerich, B., Meuser, F. and Pfaller, W. (1989) Extrusion cooking of wheat products, in *Wheat is Unique*, (ed. Y. Pomeranz) American Association of Cereal Chemists, St. Paul, Minnesota, pp. 395–419.
4. Amend, T. (1990) Mikroskopische Untersuchungen zur Teig- und Kleberbildung bei Weizen. *Dissertaton*, Technische Universitaet Muenchen, Muenchen.
5. Amend, T. and Belitz, H.-D. (1989) Microscopical studies of water/flour systems. *Z. Lebensm. Unters. Forsch.*, **189**, 103–109.
6. Meuser, F. and Smolnik, H.-D. (1979) Anwendung der Membranfiltrationstechnik zur Aufarbeitung des Prozeßwassers bei der Weizenstaerkegewinnung, in *Schriftenreihe aus dem Fachgebiet Getreidetechnologie 1*, (ed. F. Meuser) Technische Universitaet Berlin, Berlin, pp. 11–19.
7. Meuser, F., Rajani, Ch. and Juretko, A. (1977) Zur Problematik der mechanischen Trennung des Weizenendosperms in Staerke und Protein. *Getreide Mehl u Brot*, **31**, 199–204.
8. Soulaka, A. B. and Morrison, W. R. (1985) The bread baking quality of six wheat starches differing in composition and physical properties. *J. Sci. Food Agric.*, **36**, 719–727.
9. Goergens, H.-J. (1980) Siebloser Mahlautomat 'Ultra-Rotor' fuer Feinstzerkleinerung und Mahltrocknung. *Chem. Prod.*, **9**, 30–34.
10. Meuser, F. and van Lengerich, B. (1984) Systems analytical model for the extrusion of starches, in *Thermal Processing and Quality of Foods*, (eds P. Zeuthen, J. C. Cheftel, C. Eriksson, M. Jul, H. Leniger, P. Varela and G. Vos) Elsevier Applied Science, London, pp. 175–179.
11. Van Lengerich, B. (1984) Entwicklung und Anwendung eines rechnerunterstuetzten systemanalytischen Modells zur Extrusion von Staerken und staerkehaltigen Rohstoffen. *Dissertation*, Technische Universitaet Berlin, Berlin.
12. Meuser, F., van Lengerich, B. and Gimmler, N. (1990) Optimization in extrusion, in *Processing and Quality of Foods*, Vol. 1, (ed. P. Zeuthen, J. C. Cheftel, C. Eriksson, T. R. Gormley, P. Linko and K. Paulus) Elsevier Applied Science, London, pp. 1.215–1.225.
13. Pfaller, W. (1987) Einfluß der Zusammensetzung von Getriedemahlerzeugnissen und ausgewählten Rezepturen auf das Kochextrusionsverhalten und die Extrudateigenschaften. *Dissertation*, Technische Universitaet Berlin, Berlin.
14. *SAS/graph user's guide* (1981) Graph, SAS Institute, Gary, North Carolina.

14 Genetics of wheat quality and its improvement by conventional and biotechnological breeding

N. E. POGNA, R. REDAELLI, T. DACHKEVITCH, A. CURIONI and A. DAL BELIN PERUFFO

14.1 Breadmaking, pasta making and nutritional quality of wheat

Overall breadmaking quality of wheat depends on several factors which correspond to the ability to produce quality bread. The most important factors are water absorption, loaf volume, internal and external loaf characteristics, and tolerance to mixing and fermentation. All these quality factors are correlated to the physical and chemical properties of flour or dough. Several tests such as farinograph, extensograph, mixograph and alveograph can estimate the dough mixing or viscoelastic properties. Other breadmaking quality tests like the Pelshenke dough ball test, the Zeleny sedimentation test and the SDS (sodium dodecyl sulphate)-sedimentation test can give valuable information about baking quality.

Pasta-making quality generally depends upon cookability, stickiness, firmness, elasticity, cooking tolerance and water absorption. Pasta stickiness has received little attention due to difficulty in quantification.

A number of workers have developed successful methods for estimating cooked spaghetti firmness and resilience[1] and have associated cooking quality with protein content, gluten composition and solubility, farinograph mixing characteristics, SDS-sedimentation volume and mixograph characteristics.[2]

Little emphasis has been placed on improving the nutritional quality of wheat. The nutritional quality can be improved by increasing protein content and limiting amino acids especially lysine. Protein content of wheat can be significantly increased through breeding; unfortunately, a negative correlation exists between lysine expressed as a percent of protein and percent protein in common and durum wheats. Therefore lysine as percent of protein can be used as a measure of protein quality while lysine as percent of sample is a function of both the protein and lysine (percent of protein) concentrations of a sample. The inverse relationship between protein and lysine contents can be attributed to differences between albumins+ globulins and gluten. Both albumins and globulins are higher in lysine than gliadins and glutenins, whereas gliadin and glutenin are high in glutamic acid. Wheats of low-protein content usually have a higher proportion of

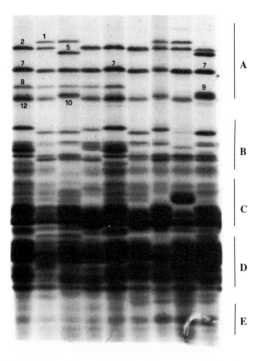

Figure 14.1 SDS-PAGE fractionation of total endosperm proteins from common wheat cultivars. The different protein fractions are indicated in the right margin. The HMW glutenin subunits have been identified by numbers. A, HMW glutenin subunits; B, ω-gliadins, HMW albumins and HMW globulins; C, LMW glutenin subunits (B subunits); D, α-, β-, γ-gliadins and LMW glutenin subunits (C subunits); E, LMW albumins.

albumins and globulins and, therefore would have a higher lysine content (as percent of protein) compared with those with higher protein content.

14.2   The endosperm proteins of wheat

Both quantity and quality of endosperm proteins are the major factors responsible for baking quality, pasta quality and nutritional value of wheat.[3] Bread flour and pasta semolina must have relatively high protein content of the right quality. Similarly, the nutritional quality of wheat used as food or feed would depend on protein quantity and on the balance of the essential amino acids, particularly lysine. Wheat endosperm protein can be divided into four solubility classes according to the classical Osborne procedure: albumins, globulins, gliadins and glutenins. However, due to unusual solubility properties and aggregative behaviour of endosperm proteins it will be more appropriate to classify on the basis of their molecular size, chemical structure, endosperm location and genetic control.[4]

Albumins and globulins, also known as 'soluble proteins', are absent from protein bodies and contain a number of proteins including enzymes, enzyme inhibitors and some inactive proteins.[5]

Gliadins and glutenins are the storage proteins and represent about 80% of the total protein in the grain; they are deposited and stored in protein bodies during endosperm development. A typical electrophoretic separation of wheat endosperm proteins is shown in Figure 14.1.

The gliadins which constitute about 40% of total endosperm protein are a heterogeneous mixture of single-chain polypeptides, alcohol-soluble in the native state and molecular weight ranging from 28 to 70 kDa.[6] Some of the gliadin molecules are stabilized by intrachain disulfide bonds. Upon electrophoresis in polyacrylamide gels at acid pH (A-PAGE), gliadins from a single variety can be separated into at least 25 protein components (Figure 14.2) designated α-, β-, γ- and ω-gliadins in order of decreasing mobility. When fractionated by a two-dimensional electrophoretic system about 50 components are separated. Because of the extensive polymorphism, gliadin patterns are widely used to check uniformity, distinctness and heterozygosity of cross-pollinated wheats.[7] Gliadins have been classified into three groups on the basis of N-terminal amino acid sequence homologies, two corresponding to ω- and γ-gliadins and the third corresponding to α- and β-gliadins combined.[8]

Gliadins have low-lysine content (0.5 mol %) and this is the main negative factor affecting the nutritional quality of wheat. Gliadins are also low in other ionic amino acids. Low levels of histidine, arginine, lysine and free carboxyl groups of glutamic acid and aspartic acid make the gliadins the least charged proteins. Glutamic and aspartic acids exist almost entirely as amides.

Two major classes of glutenin subunits have been identified in wheat endosperm: the high molecular weight (HMW) and the low molecular weight (LMW) glutenin subunits.[9] Both classes of subunits are present in the flour as cross-linked proteins resulting from interpolypeptide disulfide linkages.

The HMW glutenin subunits are a minor component in terms of quantity but are major determinants of elasticity of gluten.[10] Each common wheat cultivar possesses three to five HMW subunits (1–3 in durum wheat) which differ from the other subunits and the gliadins in MWs (90–150 kDa by SDS-PAGE) and high-glycine content.

The LMW subunits of glutenin exist as polymers stabilized by interchain disulfide bonds. They have also been called alcohol-soluble reduced glutenins, HMW gliadin, glutenin III or aggregated gliadin. Some exist as heteropolymers with HMW glutenin subunits. These heteropolymers are alcohol insoluble but soluble in dilute acid or alkali in the native (unreduced) state. After reduction of the disulfide bonds, the LMW glutenin subunits divide into two main groups; a major group of basic proteins (B subunits)

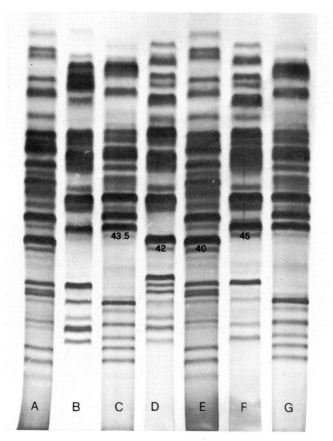

Figure 14.2 Fractionation by A-PAGE of gliadin proteins from common (lanes A, C, E and G) and durum (lanes B, D and F) wheat cultivars. The *Gli-B1* encoded γ-gliadin bands associated with gluten quality have been identified by number as described in the text (see section 14.4).

with MWs of 42–51 kDa and a minor group (C subunits) with MWs of 30–40 kDa. Because of their similar molecular size, LMW glutenin subunits are difficult to separate from α/β-gliadins and γ-gliadins by SDS-PAGE; however, they can be easily fractionated by two-dimensional electrophoretic procedures or 2-step one-dimensional electrophoresis,[11,12] and purified by combining differential solubility, ion-exchange chromatography and SDS-PAGE.[13]

14.3   Genetics of wheat endosperm proteins

Bread and durum wheats are polyploid species containing three and two related genomes, respectively. In spite of this repetition of genetic information, genetical studies indicate that endosperm protein genes exhibit simple codominant Mendelian inheritance. Research on the chromosomal location of gliadin and glutenin subunit genes and on their linkage has received great impetus from the availability of various aneuploids and partial chromosomal deletions. In some studies intervarietal substitution lines have been used as well.

14.3.1   The gliadins

Most genes coding for ω- and γ-gliadins are tightly clustered at three homologous loci names *Gli-A1* (chromosome 1A), *Gli-B1* (chromosome 1B), and *Gli-D1* (chromosome 1D) (Figure 14.3). Quantitative Southern hybridization indicated that each *Gli-1* locus contains some three to five γ-gliadin genes and a similar number of ω-gliadin genes. Recombination is extremely rare among genes at each locus. To the best of our knowledge, only two instances have been reported. One of these occurred at the *Gli-B1* locus during the breeding of the durum wheat cultivar Berillo.[14] Because of the close linkage, the gliadin polypeptides coded by each locus are inherited strictly as a block, which is referred to as a gliadin allele. Multiple codominant alleles are known for each locus; it is also likely that different alleles contain different numbers of active genes.

A new catalogue of alleles at *Gli-1* loci based on the analysis of 45 crosses and about 360 bread-wheat cultivars from seven countries includes 18 alleles

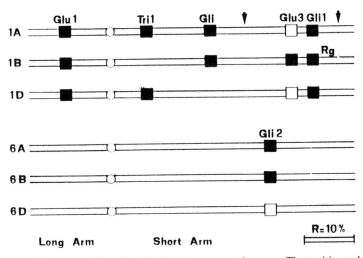

Figure 14.3 Chromosomal location of wheat storage protein genes. The positions which are depicted by open squares have not been determined. Only parts of the long arm are shown. Arrow heads indicate the positions of dispersed genes coding for minor ω-gliadin bands. R = recombination frequency, Rg = red glume locus, Tri1 = Triplet proteins, Gli = gliadin loci, Glu = glutenin subunit loci.

for the *Gli-A1* locus, 16 for the *Gli-B1* locus and 12 for the *Gli-D1* locus.[15] Several alleles in each locus are likely to originate through intralocus recombination. Some alleles such as *Gli-A1i* and *Gli-A1j* may originate from each other as a result of point mutation affecting the electrophoretic mobility of one gliadin polypeptide, whereas the *Gli-B1$_j$* allele could arise from the *Gli-B1b* allele through mutational switching-off or deletion of one gliadin gene. The polymorphism of gliadins is considerably higher than that reported in the above mentioned catalogue. Many new alleles were found in other hexaploid wheats such as *Triticum macha* or in tetraploid and diploid wheats.[15]

The *Gli-1* loci have been shown to occur at the same relative position towards the end of the short arms of the three homologous chromosomes; in particular the *Gli-B1* locus is physically located close to the distal end of the chromosome 1B satellite.[16,17] The order of ω- and γ-gliadins genes within the *Gli-B1* locus with respect to the centromere has been recently inferred from the recombination data of crosses involving the durum wheat cultivar Berillo.[14]

A few ω-gliadins are coded by additional dispersed genes which are remote from the main clusters and which are able to recombine with them.[18] Some of these 'selfish' genes occur on the short arms of chromosome 1A and 1B at new loci, about 30 recombination units from *Gli-1*. Two additional ω-gliadins genes are located on both sides of the *Gli-A1* locus and recombine with it at frequencies of 5% and 13%. Recombination frequencies of 1–2% have been recently obtained between *Gli-B1* and a single dispersed gene controlling the synthesis of one ω-gliadin band in the Italian cultivar Salmone; this 'selfish' gene is located proximal to the glume colour *Rg1* gene on the satellite of chromosome 1B (Figure 14.3).

When F2 grains from the cross between the Italian cultivar Costantino and the Canadian cultivar Neepawa were analysed, three major 'selfish' gliadin components were detected in the chromosome 1B of the cultivar Neepawa (Figure 14.4).

Because a number of cultivars contain biotypes which differ only by the presence or absence of such 'selfish' ω-gliadin components,[18] the occurrence of these removed genes is probably a distinctive feature of some wheat genotypes. Their effects on gluten viscoelastic properties remain unknown.

The organization of the family of gliadin genes in the group 1 chromosomes is analogous to the situation with hordein-coding genes located on chromosome 5 ($\approx$ 1H) in barley. Also, in maize, the zein-controlling genes are distributed over several chromosomes and some of them are linked in a tight cluster. Therefore it seems that the presence of clusters and removed single genes on the same chromosome is a general characteristic of prolamin genes.

Another distinctive feature of the *Gli-1* loci is their linkage with genes for resistance to fungal diseases (powdery mildew, *Pm3* locus; leaf rust, *Lr10*

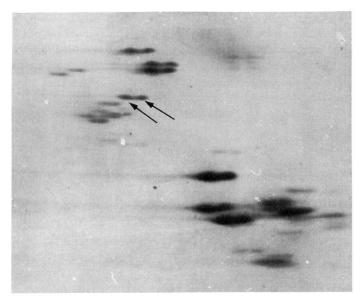

Figure 14.4 Two-dimensional fractionation of gliadins of cv. Neepawa. Gliadins coded by 'selfish' genes on chromosome 1AS are marked with arrows.

locus; yellow rust, *Yr10* locus) and with genes controlling morphological characters of the plant such as glume hairiness (*Hg1* locus) and glume colour (*Rg1* locus). Further investigations on the arrangements of the gliadin genes and the association of specific gliadin polypeptides with quality or other agronomic characteristics will be facilitated by the use of these genes as 'flanking' markers. On the other hand, gliadin genes can be used as reporter genes for introgression of resistance genes into susceptible cultivars.

DNA sequencing of gliadin genes has revealed that α- and β-gliadin genes are tightly clustered at a single locus on each of the chromosomes of group 6 (Figure 14.3). The *Gli-2* loci are located towards the distal ends of the short arms; in particular *Gli-B2* occurs on the satellite segment of chromosome 6B. Each locus codes for five to ten major proteins and no recombination with a *Gli-2* locus has been reported.[19] According to Payne,[19] genes at the *Gli-1* and *Gli-2* loci should arise from duplication and divergence of a common ancestral gene, the spatial separation of these loci probably being due to an ancient interchromosomal translocation that occurred prior to evolution of the ancestral diploid wheat species.

There are at least 24 *Gli-A2* alleles (chromosome 6A), 22 *Gli-B2* alleles (chromosome 6B) and 19 *Gli-D2* alleles (chromosome 6D).[15] Most of them originate from one another through single mutation events. In total, 111 gliadin alleles have been recognized, which corresponds to $18 \times 16 \times 12 \times 24 \times 22 \times 19 \cong 34 \times 10^6$ wheat genotypes. Indeed, this great variation is exploited to identify varieties and varietal mixtures of seed or flour, to distinguish between species or between genera, to investigate pedigrees and for a range of applications in breeding studies. The same chromosomal location of genes coding for gliadins was shown in durum wheats using a set of D-genome disomic-substitution lines in which the seven D chromosomes were individually substituted for their respective A or B homologues.[20]

14.3.2   High molecular weight (HMW) glutenin subunits

It has been known for some time that the genes coding for the HMW glutenin subunits (also referred to as A subunits) are located on the long arm of chromosomes 1A, 1B and 1D[19] (Figure 14.3). Each of the three loci, collectively named *Glu-1*, contains only two genes names *Glu-1-1* and *Glu-1-2*, which code for *x*-type and *y*-type subunits, respectively. Individual cultivars of *T. aestivum* have 0 or 1 subunit controlled by *Glu-A1* (chromosome 1A), 1 or 2 subunits controlled by *Glu-B1* (chromosome 1B) and 2 subunits controlled by *Glu-D1* (chromosome 1D). Intralocus recombination is very rare; map distance between the two genes of the *Glu-B1* locus has been estimated at 0.1–0.2 cM.[19] The HMW glutenin subunit genes exhibit extensive allelic variation and many alleles have been recently identified in both common and durum wheats and in diploid species.[21]

For example, we have recently found (unpublished) a novel HMW subunit pair in a French wheat line (Figure 14.5). The *y*-type subunit of this pair has a much higher electrophoretic mobility in SDS-PAGE, and therefore probably a much smaller size, than any other HMW subunit found in hexaploid wheat. Analysis of F2 segregation data suggests that the synthesis of this subunit pair is controlled by the *Glu-D1* locus. The effects of this new allele on breadmaking quality and other characters are currently being tested on F4 segregating progeny.

The *Glu-1* loci segregate independently from *Gli-1* loci. The recombination frequencies between *Glu-1* and the centromeres obtained from different types of recombination analysis range from about 10–27.5%.[22] Two chromosome mutants lacking half of the long arm of chromosomes 1B or 1D as well as the HMW subunits encoded by these chromosomes have been described by Payne[19] suggesting that the *Glu-1* loci are located on the distal half of the long arms. The *Glu-A1-2* gene does not code for any HMW subunit in bread-wheat cultivars but is active in some diploid wheats that contain A-type genomes. Cultivars showing silent *Glu-A1-1* and *Glu-B1-2* genes are known as well. Recently a common wheat line lacking all HMW glutenin subunits has been produced.[23]

There is little variation in the amount of individual subunits synthesized when alleles of the same locus are compared. However, a genotype from Israel, TAA 36, was found to overproduce a 1Bx subunit, compared to other HMW subunits of glutenin.[19] Also, three Canadian hard red spring wheat

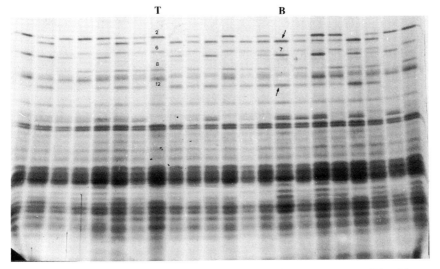

Figure 14.5 SDS-PAGE of total proteins of F1 single seeds from the cross Ben (B) × Thesee (T). The novel *Glu-D1* encoded HMW subunits in cv. Ben are indicated by arrows.

cultivars, Roblin, Bluesky and Glenlea, which have very strong gluten, overproduce the *x*-type subunit 7 controlled by the *Glu-B1b* allele.[24] This may be explained by duplication of the *Glu-B1-1* gene or by a more efficient transcription of the gene.

Knowledge about HMW subunit composition of the different wheat cultivars is of great importance because there is strong evidence that the presence of certain HMW subunits is positively correlated with improved breadmaking quality. This aspect will be considered later.

14.3.3   Low molecular weight (LMW) glutenin subunits

In contrast to gliadins and HMW glutenin subunits which are easily resolved by A-PAGE or 1-D SDS-PAGE, the LMW glutenin subunits (also known as B and C subunits) have proved much more difficult to analyse by 1D-SDS-PAGE because of overlap with the gliadins. The introduction of a two-step 1-D SDS-PAGE or Acid-PAGE/SDS-PAGE procedure provided a rapid method for analysing a large number of samples in a gliadin-free background. Genetic evidence has indicated that the LMW glutenin subunits are controlled by genes at *Glu-A3*, *Glu-B3* and *Glu-D3* loci on the chromosome arms 1AS, 1BS and 1DS, respectively. Amongst a collection of 222 hexaploid wheats from 32 countries, 20 band patterns (LMW glutenin blocks) were detected, six for the *Glu-A3* locus, nine for the *Glu-B3* locus and five for the *Glu-D3* locus.[25] Chromosome 1A encodes for a few LMW subunits and many cultivars do not exhibit any LMW subunit coded by

Glu-A3, whereas a great polymorphism exists for subunits encoded by chromosome 1B.

Genes coding for LMW subunits are closely linked. Moreover, LMW subunit genes have been found to be linked to genes coding for ω- and γ-gliadins. The estimated map distance is 2 cM between *Glu-B3* and *Gli-B1* on the short arm of chromosome 1B in both bread and durum wheat. Recent results[26,27] also suggest that *Glu-B3* is located between *Glu-B1* and *Gli-B1* (Figure 14.3). There are no recombination data available for LMW subunits coded by chromosomes 1A and 1D, but a similar, relative position is anticipated. Allelic variations in the LMW subunits have recently been shown to be primarily responsible for differences in gluten viscoelastic properties in both bread and durum wheats.[14,28]

14.3.4   The D-zone omega gliadin

Some gliadins have been found to be strongly associated with glutenins without covalent linkage.[29] Some of these gliadins belong to the ω-gliadin group and upon SDS-PAGE fall in the molecular weight zone of 40–60 kDa, called the D zone. The D-zone gliadins are controlled by genes on the chromosome arms 1AS, 1BS, 1DS at the *Gli-A3*, *Gli-B3* and *Gli-D3* loci, respectively.[30] These loci seem to be closely linked to the gliadin coding loci *Gli-1*. Because of their aggregative behaviour, the D-zone ω-gliadins may affect technological qualities of wheat flours.

14.3.5   The soluble proteins

The albumin–globulin fraction, which represents 15–25% of wheat endosperm proteins, includes the CM proteins (chloroform-methanol-soluble), many α-amylase/trypsin inhibitors, the HMW albumins, the 'triplet protein' and several other non-storage proteins. Some of these 'soluble proteins' have been detected in gluten prepared by washing starch from a dough.[31] For example, the low-molecular weight gluten fraction, named S protein, was also found in flour albumins/globulins prepared by Osborne fractionation.[32] Soluble proteins affect breadmaking or pasta making quality. The amount of a highly aggregating albumin fraction was shown to be inversely correlated to breadmaking quality.[33] Moroever, it was suggested that S proteins may be important in breadmaking quality because they have high affinity for polar lipids, which may positively affect loaf volume.[31] In durum wheat, the surface state of cooked pasta was found to be related to a sulfur-rich protein fraction, termed DSG proteins, which belongs to the CM protein family.[34]

CM proteins are comprised of three main components (CM1, CM2 and CM3) and 2–3 minor proteins designated 11, 16 and 17.[35] They have molecular weights in the range of 11–13 kDa and unique amino acid compositions with high levels of lysine and non-polar amino acids. The CM proteins are encoded by genes located in chromosomes of groups 4 and 7; the CM-1 protein by chromosome 7DS, CM-2 by chromosome 7BS, CM-3 by chromosome 4A and by an unknown chromosome of the D genome.

'Triplet proteins', now renamed 'triticins', are minor endosperm proteins deposited in protein bodies which appear as a triplet of HMW bands (called Tri-1, Tri-2 and Tri-3) in non-reduced fractionations by SDS-PAGE.[36,37] Upon reduction, the proteins appear to be heterotetramers of subunits designated D (MW 58 kDa), δ (22 kDa), A (52 kDa) and α (23 kDa) where Tri-1 = D.δ.D.δ., Tri-2 = D.δ.A.α and Tri-3 = A.α.A.α. Triticin polypeptides occur in wheat endosperm as HMW aggregates in the glutenin complex held together by hydrophobic or hydrogen bonds. They belong to the same family of proteins as the 11-12S globulin storage proteins of legumes, oats and rice, and have higher amounts of lysine, threonine and sulfur amino acids than the gluten proteins.

The small and large subunits of triticin proteins are controlled by two loci called *Tri-A1* (subunits A and α) and *Tri-D1* (D and δ) on the short arm of chromosomes 1A and 1D. No homoallele on the short arm of chromosome 1B has yet been found. The *Tri-A1* and *Tri-D1* loci are closely linked to the centromere on chromosome arms 1AS and 1DS and loosely linked to the gliadin genes *Gli-1*.

Numerous protein components are present in SDS-PAGE fractionation of albumins from bread-wheat flour (Figure 14.6). The major proteins in this fraction occur in two different size groups; the high molecular weight (HMW) albumins (about 60 kDa) and the low molecular weight (LMW) albumins (14–16 kDa). Certain HMW albumins bands (MW 65, 63, 60 kDa) which occur in both disulfide-linked oligomers and monomeric forms in their native state, are the β-amylases controlled by chromosome arms 4DL, 4AL and 5AL (β-*Amy-1* loci). A limited amount of allelic variation in these bands has been observed. Because bands with mobilities similar to β-amylases have been seen in abundant amounts in a protein fraction imparting poor breadmaking properties to doughs, these HMW albumins could be of significant importance to wheat breeders.

A major component of the LMW albumin proteins was found to be encoded by the short arm of chromosome 3D, as suggested by its absence in the water extracts of aneuploid strains nulli 3D-tetra 3B and ditelo 3DL of Chinese Spring.[38] The antiserum to this component has been found to interact with several members of the CM-protein family.[38]

14.4 Protein composition and gluten quality for breadmaking

Recognition of the heterogeneity of gliadins and glutenins has permitted a new approach in the analysis of the biochemical and genetic factors involved

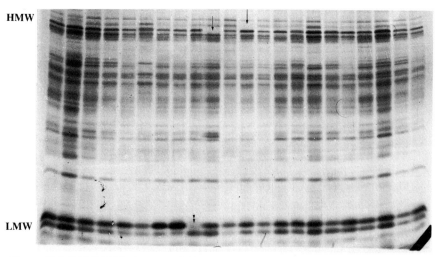

Figure 14.6 SDS-PAGE of water-soluble endosperm proteins from aneuploid lines of cv. Chinese Spring. HMW albumins coded by chromosome 4AL and 4DL are marked with arrows. LMW albumins coded by chromosome 3DS are marked by arrowheads.

in wheat quality. Gliadins are generally considered to contribute to the viscosity and extensibility of gluten whereas glutenins appear to be the major determinant of elasticity. Significant correlations between gliadin electrophoretic components and gluten quality have been observed by several authors. For example, a consistent relationship was found between the presence of the γ-gliadin designated band 45 and gluten strength, and between the presence of another γ-gliadin, band 42, and gluten weakness in durum wheats.[39] These γ-gliadins are allelic forms of a protein encoded by the gene locus *Gli-B1* on chromosome 1B and are present in durum wheat cultivars from diverse sources (Figure 14.2). Several authors confirmed that cultivars or segregating lines with band 45 (and other linked ω-gliadins) have higher SDS-sedimentation values, cooking or overcooking quality and mixograph mixing times than those with band 42. Results of analyses carried out in our laboratories indicated that γ-gliadin 45 has a strongly favourable influence on the breadmaking quality of durum wheats as well.[40] The genetic linkage between gliadins 42/45 and the glume colour character controlled by the *Rg1* gene locus (the dominant and recessive alleles determine brown and white colours, respectively) is now exploited in screening F3 progenies for gluten strength.

Recently, gliadins 42 and 45 have been found to be only genetic markers of quality, whereas allelic variation for LMW glutenin subunits encoded at the *Glu-B3* locus is primarily responsible for differences in gluten viscoelastic properties.[14] In bread-wheat cultivars, there are two allelic γ-gliadins, designated 43.5 and 40, which are quite similar to bands 45 and 42 in electrophoretic mobilities, molecular weights, amino acid composition and viscoelastic properties, γ-gliadin 43.5 having a favourable effect on bread-making quality (Figure 14.2). As for γ-gliadins 45 and 42, genes for bands 43.5 and 40 are linked tightly to different blocks of ω-gliadins and LMW glutenin subunits.[41]

Several HMW subunits of glutenin are associated with breadmaking quality. Payne *et al.*[19] have analysed numerous unselected progeny of crosses between common wheat cultivars for both SDS-sedimentation volume (which is correlated with loaf volume) and subunit composition and showed that certain allelic subunits impart differential effects on gluten quality. Since the *x*- and *y*-type of HMW subunits encoded by the *Glu-B1* and *Blu-D1* loci are always inherited as pairs, it was not possible to determine the effects of individual subunits on gluten quality. However, analysis of the segregating progeny from crosses involving the recombinant cultivar Fiorello, which contains the unusual combination of subunits 1Dx5 and 1Dy12, suggested that the *y*-type subunit 10 is responsible for the 'good quality' of the subunit pair 5 + 10.[42] Isolation and technological analysis of further recombinant lines will allow establishment of the effects of single subunits and of their interactions on gluten viscoelastic properties. Also, some *Glu-B1* alleles have been found to affect breadmaking and pasta-making quality of durum wheat cultivars. These results have been confirmed by several authors. Branlard and Dardevet,[43] for example, have shown that the alveograph parameters *W* (gluten strength) and *P* (tenacity) and the Zeleny sedimentation value are correlated positively with subunits 7 + 9 and 5 + 10, and negatively with bands 2 + 12, whereas subunit 1 is correlated with *W*, and subunits 2* and 17 + 18 with *G* (swelling).

Allelic variation in LMW and HMW glutenin subunits has cumulative effects on dough properties. Some LMW glutenin alleles have been found to be correlated with dough extensibility.[44] Although the protein content of flour has a large effect on dough extensibility compared to the LMW glutenin alleles, the selection based on glutenin allele composition can help breeders develop new wheat lines with large dough extensibility even at relatively low-protein levels.

14.5 Protein composition and nutritional quality

Lysine is the first limiting amino acid in wheat and its low content is a direct consequence of the high content of gliadins. Albumins and globulins, present in low proportion in wheat, have an amino acid composition that fits the requirements of humans and monogastric animals. The glutenin fraction has an intermediate lysine content. Therefore attempts to improve the nutritive values of wheat proteins should concentrate on altering the content or composition of the gliadins. This could be obtained (i) by changing the

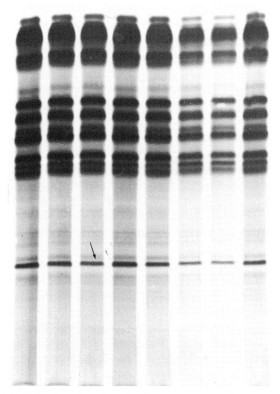

Figure 14.7 A-PAGE fractionation of gliadins from bread-wheat lines lacking gliadins coded by the *Gli-B1* and *Gli-D1* loci. The *Gli-A1* encoded ω-gliadin band is arrowed.

amino acid composition of gliadins, (ii) by increasing the content of storage proteins with a better nutritional value or (iii) by decreasing the gliadin content. Genes for significantly higher-than-normal lysine (as percentage of protein) are well known in maize, sorghum and barley. In wheat, varieties or mutants containing significantly higher lysine content in endosperm protein have not been recorded, probably because the polyploidy makes the detection of recessive mutants much more difficult. A possible way to improve the composition of gliadins can result from the large genetic variation in the composition of gliadins which show some variability in lysine content.[45] However, the storage protein genes consist of multigene families, a fact that makes the combination of genes coding for better polypeptides very difficult. An alternative approach would be to exploit the polyploid nature of commercially important wheats for replacing gliadins by more lysine-rich proteins such as the albumins and the globulins. In fact various chromosome deletions and duplications can be compensated for by a normal genome in tetraploids and hexaploids, as is clearly shown by the fertility and the viability of most ditelosomic and addition lines in common wheat. As previously shown, the *Gli-1* loci occur near the ends of chromosomes so that a chromosome break may delete the whole locus. Spontaneous mutants with deletion of either the 1B satellite or a very small piece of chromosome containing the *Gli-D1* locus have been isolated in commercial cultivars of common wheat. In these deletion lines fertility is not reduced; they were used in crosses to produce double mutants which lack both 1B- and 1D-encoded gliadins and the LMW glutenin subunits (Figure 14.7). The double mutants which have normal seed set and protein content, are being tested for endosperm protein composition and quality. New nulls for gliadins coded by *Gli-A1*, *Gli-A2*, *Gli-B2* and *Gli-D2* and new variant forms producing very low levels of gliadins coded by *Gli-B2* or *Gli-D2* have been found in commercial cultivars or landraces of both common and durum wheats and used in crosses to produce multiple nulls. These lines are now being investigated as a possible approach to increasing the lysine content of wheat.

As previously indicated, a significant proportion of endosperm protein consists of 'soluble proteins' (albumins and globulins) which have a more balanced amino acid composition. For example, triticin polypeptides have balanced amounts of lysine and threonine compared to gliadins and glutenin subunits.[37] Genetic manipulation of genes coding for triticin may be another way of improving the nutritional quality of wheat.

14.6 New technologies for the improvement of wheat quality

14.6.1 In vitro *tissue culture*

New biotechnologies which operate at the cellular level are based on the observation that single cells, protoplasts and microspores can regenerate new plants when grown *in vitro* on artificial media. Immature embryos and anthers of wheat are the most efficient sources of tissue for regenerating whole plants in large numbers from calli. Some of these regenerated plants show genetic variations (e.g. somaclonal variation) that are apparently the result of genetic changes that occur during the time the calli are in culture. Somaclonal variants have been obtained in many cereals for some traits of agronomic importance.[46] Somaclonal variations for storage protein compositions could be a new and powerful source of genetic variation for breeding for technological and nutritional quality.

Anther culture is potentially the most efficient haploid production technique in bread wheat. In fact, androgenesis is used widely in plant breeding laboratories because of its ability to achieve homozygosity rapidly compared with conventional breeding. A few commercial cultivars have already been developed in France and China.[47,48] Moreover, protoplasts or calli developed from anthers may be the best targets for introduction of foreign genes into wheat because of their haploid constitution.

In recent years a large number of bread wheat cultivars have been evaluated for embryoid induction rate and frequency of green plant regeneration. Genetic analyses indicate that callus and green plant production are two independently inherited characters controlled by a low number of 'major genes'. Genes on chromosomes 1D and 5B have been found to increase embryoid frequency. Moreover, the significant promotion of callus development and green plant regeneration induced by the 1BL/1RS translocation present in some bread-wheat cultivars indicates that the 1RS chromosome contains androgenetic gene(s).

In durum wheat, the low level of pollen callus induction and the extremely low yield, if any, of green plantlets have hampered the successful application of androgenesis in plant breeding and genetic research. However, we have recently analysed a durum wheat line homozygous for the 1BL/1RS translocation produced by crossing the durum wheat cultivar Cando with the 1BL/1RS translocation wheat cv Veery.[49] Embryoid production frequency in this durum line was similar to that of the bread-wheat parent Veery and much higher than those of the durum wheat cultivars.[50] About 17% of the embryoids from the translocation durum line regenerated plantlets and two out of 21 of them were green. Since the parent Veery had a high regeneration ability and a high proportion of green plant compared with the 1BL/1RS translocation durum line, the chromosome arm 1RS must have an enhancing effect on embryoid induction without affecting the rate of green plant regeneration.

14.6.2   Gene transfer in wheat

There have been several reports of introduction of foreign genes into dicotyledonous plants in such a way that they are expressed in the recipient plant. In the wheat species, transgenic plants are not yet available. However, two new techniques for plant gene transfer are being currently investigated, the 'biolistic process' and the 'PEG-mediated DNA transfer'. The biolistic process[51] relies on the acceleration to high velocity of small high-density metal particles (tungsten or gold microprojectiles), so that they can penetrate the cell wall and membrane of the target cells and deliver foreign DNA which had been previously adsorbed onto their surface. Cell suspension cultures, callus cultures and intact tissues such as meristems and organs can be used as target. Some species such as tobacco, soybean, maize and cotton, have been stably transformed by the biolistic process.[52–54]

The direct gene transfer using PEG (polyethylene glycol) relies on the treatment of embryogenic protoplasts with PEG in the presence of foreign DNA. Recently, methods for protoplast isolation and for plant regeneration from suspension culture in wheat have been developed. Therefore there is good reason to be optimistic that transgenic wheat plants will be obtained over the next few years.

At the present time the amount of information concerning the function of wheat endosperm proteins and their genes is sufficient to assess how the introduction of cloned genes and the amino acid replacement substitutions might improve dough and nutritional qualities. For example, the introgression of HMW glutenin subunit genes by transformation will be more rapid than by a conventional backcrossing procedure. Moreover, the regulatory regions of storage protein genes may be manipulated before gene insertion in order to enhance the effects of the 'good' alleles on dough quality. Finally, nutritionally balanced storage protein genes may be constructed and introduced by molecular transformation to increase the proportion of lysine and threonine.

14.6.3 Hybrid wheats

Following the discovery of effect chemical hybridizing agents (CHA), thousands of F1 bread wheat hybrids have been produced and tested in several wheat growing areas of the world. Recently, information on breadmaking quality and on the combining ability for some technological parameters related to breadmaking quality has come from analysis of 200 F1 grown in different locations.[55] In general, hybrid wheat tends to have high protein content and enhanced dough extensibility compared with the parents. Significant positive general combining ability effects were observed for yield, protein content and gluten viscoelastic properties.

Although information on the economic implications of the introduction of hybrid wheat into practical agriculture is needed, yet, based on the storage protein composition of the parents, it seems that F1 hybrid wheats can have satisfactory breadmaking properties and high-yield potential.

References

1. Dexter, J. E., Matsuo, R. R. and MacGregor, A. W. (1985) Relationship of instrumental assessment of spaghetti cooking quality to the type and the amount of material rinsed from cooked spaghetti. *J. Cereal Sci.*, **3**, 39–53.
2. Matsuo, R. R., Dexter, J. E., Kosmolak, F. G. and Leisle, D. (1982) Statistical evaluation of tests for assessing spaghetti-making quality of durum wheat. *Cereal Chem.*, **59**, 222–228.
3. Orth, R. A. and Bushuk, W. (1972) A comparative study of the proteins of wheats of diverse baking qualities. *Cereal Chem.*, **49**, 168–175.
4. Miflin, J. B., Field, J. M. and Shewry, P. R. (1983) Cereal storage proteins and their effect on technological properties, in *Phytechemical Society of Europe Symposium, Volume 20: Seed proteins*, (ed. J. Daussant, J. Moose and J. Vaughan) Academic Press, New York, pp. 255–319.
5. Payne, P. I., Holt, L. M., Jaruis, M. G. and Jackson, E. A. (1985) Two-dimensional fractionation of the endosperm proteins of bread wheat (*T. aestivum*): biochemical and genetic studies. *Cereal Chem.*, **62**, 319–326.
6. Payne, P. I., Holt, L. M., Lawrence, G. J. and Law, C. N. (1982) The genetics of gliadin

and glutenin, the major storage proteins of the wheat endosperm. *Qual. Plant. Foods Hum.*, **31**, 229–241.

7. Pogna, N. E., Borghi, B., Mellini, F., Dal Belin Peruffo, A. and Nash, R. J. (1986) Electrophoresis of gliadins for estimating the genetic purity in hybrid wheat seed production. *Genet. Agric.*, **40(2)**, 205–212.

8. Kasarda, D. D., Autran, J-C., Lew, E.J-L., Nimmo, C. C. and Shewry, P. R. (1982) N-Terminal amino acid sequence of ω-gliadins and ω-secalins. Implications for the evolution of prolamin genes. *Biochim. Biophys. Acta*, **747**, 138–150.

9. Huebner, F. R. and Wall, J. S. (1976) Fractionation and quantitative differences of glutenin from wheat varieties varying in baking quality. *Cereal Chem.*, **53**, 258–263.

10. Payne, P. I., Harris, P. A., Law, C. N., Holt, L. M. and Blackman, J. A. (1980) The high-molecular-weight subunits of glutenin: Structure, genetics and relationships to bread-making quality. *Ann. Technol. Agric.*, **29(2)**, 309–320.

11. Jackson, E. A., Holt, L. M. and Payne, P. I. (1983) Characterization of high-molecular-weight gliadin and low-molecular-weight glutenin subunits of wheat endosperm by two-dimensional electrophoresis and the chromosomal localization of their controlling genes. *Theor. Appl. Genet.*, **66**, 29–37.

12. Singh, N. K. and Shepherd, K. W. (1988) Linkage mapping of the genes controlling endosperm proteins in wheat. 1. Genes on the short arms of group 1 chromosomes. *Theor. Appl. Genet.*, **75**, 628–641.

13. Autran, J-C., Laignelet, B. and Morel, M. H. (1987) Characterisation and quantification of low-molecular-weight glutenins in durum wheats. *Biochimie*, **69**, 129–144.

14. Pogna, N. E., Autran, J-C., Mellini, F., Lafiandra, D. and Feillet, P. (1990) Chromosome 1B-encoded gliadins and glutenin subunits in durum wheat: Genetics and relationship to gluten strength. *J. Cereal Sci.*, **11**, 15–34.

15. Metakovsky, E. V. (1991) Gliadin allele identification in common wheat. II. Catalogue of gliadin alleles in common wheat. *J. Genet. Breeding*, **45**, 325–344.

16. Payne, P. I., Holt, L. M., Hutchinson, J. and Bennet, M. D. (1984) Development and characterisation of a line of bread wheat, *T. aestivum*, which lacks the short arm satellite of chromosome 1B and the *Gli-B1* locus. *Theor. Appl. Genet.*, **68(4)**, 327–334.

17. Pogna, N. E., Dal Belin Peruffo, A. and Mellini, F. (1985) Genetic aspects of gliadin bands 40 and 43.5 associated with gluten strength. *Genet. Agric.*, **39**, 101–108.

18. Metakovsky, E. V., Ackmedov, M. G. and Sozinov, A. A. (1986) Genetic analysis of gliadin-encoding genes reveals gene clusters as well as single remote genes. *Theor. Appl. Genet.*, **73(2)**, 278–285.

19. Payne, P. I. (1987) Genetics of wheat storage proteins and the effect of allelic variation on bread-making quality. *Ann. Rev. Plant Physiol.*, **38**, 141–153.

20. Du Cros, D. L., Joppa, L. R. and Wrigley, C. W. (1983) Two-dimensional analysis of gliadin proteins associated with quality in durum wheat: Chromosomal location of genes for their synthesis. *Theor. Appl. Genet.*, **66**, 297–302.

21. McIntosh, R. A., Hart, G. E. and Gale, M. D. (1990) Catalogue of gene symbols for wheat. *1990 Supplement Cereal Res. Comm.*, **18**, 141–157.

22. Shewry, P. R., Halford, N. G. and Tatham, A. S. (1989) The high molecular weight subunits of wheat, barley and rye: genetics, molecular biology, chemistry and role in wheat gluten structure and functionality. *Oxford Surveys Plant Molec. Cell Biol.*, **6**, 163–219.

23. Lawrence, G. J., MacRitchie, F. and Wrigley, C. W. (1988) Dough and baking quality of wheat lines deficient in glutenin subunits controlled by the *Glu-A1*, *Glu-B1* and *Glu-D1* loci. *J. Cereal Sci.*, **7(2)**, 109–112.

24. Ng, P. K. W., Pogna, N. E., Mellini, F. and Bushuk, W. (1989) *Glu-1* allele composition of the wheat cultivars registered in Canada. *J. Genet. Breeding*, **43**, 53–59.

25. Gupta, R. B. and Shepherd, K. W. (1990) One-dimensional SDS-PAGE analysis of LMW subunits in hexaploid wheats. *Theor. Appl. Genet.*, **80**, 65–74.

26. Gupta, R. B. and Shepherd, K. W. (1988) Low-molecular-weight glutenin subunits in wheat: their variation, inheritance and association with physical dough properties, in *Proceedings 7th International Wheat Genetics Symposium*, Cambridge University Press, Cambridge, pp. 943–949.

27. Pogna, N. E. and Mellini, F. (1988) Wheat storage protein genes and their use for improvement of pasta-making quality, in *Proceedings 3rd International Symposium on Durum Wheat*, (ed. G. Wittmer) Chamber of Commerce of Foggia, pp. 135–146.

28. Gupta, R. B., Bekes, F. and Wrigley, C. W. (1991) Prediction of physical dough properties from glutenin subunit composition in bread wheats: correlation studies. *Cereal Chem.*, **68**, 328–333.
29. Bietz, J. A. and Wall, J. S. (1975) The effect of various extractants on the subunit composition and association of wheat glutenins. *Cereal Chem.*, **52**, 145–155.
30. Payne, P. I., Holt, L. M. and Lister, P. (1988) *Gli-A3* and *Gli-B3*, two newly designated coding for some omega-type gliadins and D subunits of glutenins, in *Proceedings 7th International Wheat Genetics Symposium*, (Vol. 2), Cambridge, Cambridge University Press, pp. 999–1002.
31. Zawistowka, U., Bietz, J. A. and Bushuk, W. (1986) Characterization of low molecular weight protein with high affinity for flour lipid from two wheat classes. *Cereal Chem.*, **63**, 414–419.
32. Zawistowska, U., Bekes, F. and Bushuk, W. (1985) Gluten proteins with high affinity to flour lipids. *Cereal Chem.*, **62(4)**, 248–289.
33. Arakawa, T., Yoshida, M., Morishita, M., Honda, J. and Yonezawa, U. (1977) Relation between aggregation of glutenin and its polypeptide compositon. *Agric. Biol. Chem.*, **41**, 995–999.
34. Kobrehel, K. and Alary, R. (1989) The role of a low molecular weight glutenin fraction in the cooking quality of durum wheat pasta. *J. Sci. Food Agric.*, **47**, 487–500.
35. Salcedo, G., Fra-Mon, P., Molina-Cano, J. L., Aragoncillo, C. and Garcia-Olmedo, F. (1984) Genetics of CM-proteins (A-hordeins) in barley. *Theor. Appl. Genet.*, **68(1/2)**, 53–59.
36. Singh, N. K. and Shepherd, K. W. (1985) The structure and genetic control of a new class of disulphide-linked proteins in wheat endosperm. *Theor. Appl. Genet.*, **71(1)**, 79–92.
37. Singh, N. K., Shepherd, K. W., Langridge, P. and Gruen, L. C. (1991) Purification and biochemical characterization of triticin, a legumin-like protein in wheat endosperm. *J. Cereal Sci.*, **13**, 207–220.
38. Pogna, N. E., Redaelli, R., Beretta, A. M., Curioni, A. and Dal Belin Peruffo, A. (1991) The water-soluble proteins of wheat: biochemical and immunological studies. In *Gluten Proteins*, (ed. W. Bushuk and R. Tkachuk) AACC, St. Paul, Minnesota, pp. 407–413.
39. Damidaux, R., Autran, J-C., Grignac, P. and Feillet, P. (1978) Evidence of relationships useful for breeding between the electrophoretic pattern of gliadins and the viscoelastic properties of the gluten in *Triticum durum* Dest. *C.R. Acad. Sci. Paris, Serie D*, **287(7)**, 701–704.
40. Boggini, G. and Pogna, N. E. (1989) The breadmaking quality and storage protein composition of Italian durum wheat. *J. Cereal Sci.*, **9**, 131–138.
41. Dal Belin Peruffo, A., Pogna, N. E., Tealdo, E., Tutta, C. and Albuzio, A. (1985) Isolation and partial characterization of γ-gliadins 40 and 43.5 associated with quality in common wheat. *J. Cereal Sci.*, **3**, 355–362.
42. Pogna, N. E., Mellini, F. and Dal Belin Peruffo, A. (1987) Glutenin subunits of Italian common wheats of good breadmaking quality and comparative effects of high molecular weight glutenin subunits 2 and 5, 10 and 12 on flour quality, in *Hard Wheat: Agronomic, Technological, Biochemical and Genetic Aspects*, (ed. B. Borghi) CEC publication EUR 11172 EN, pp. 53–69.
43. Branlard, G. and Dardevet, M. (1985) Diversity of grain protein and bread wheat quality. II. Correlation between HMW subunits of glutenin and flour quality characteristics. *J. Cereal Sci.*, **3**, 345–354.
44. Gupta, R. B., Singh, N. K. and Shepherd, K.W. (1989) The cumulative effect of allelic variation of LMW and HMW glutenin subunits on dough properties in the progeny of two bread wheats. *Theor. Appl. Genet.*, **77**, 57–64.
45. Okita, T. W., Cheesbrough, V. and Reeves, C. S. (1985) Evolution and heterogeneity of the α/β-type and γ-type gliadin DNA sequences. *J. Biol. Chem.*, **260**, 8203–8213.
46. Blight, S., Oomo, C., Foulger, D. and Karp, A. (1986) Mutation and tissue culture. In *Plant Tissue Culture and Its Agricultural Application*. (ed. L. Withers and P. Alderson) Butterworths, London, pp. 431–449.
47. De Buyser, J., Henry, Y., Lonnet, P., Hertzog, R. and Hespel, A. (1987) 'Florin': A doubled haploid wheat variety developed by the anther culture method. *Plant Breeding*, **98**, 53–56.
48. Hu, D., Tang, Y., Yuan, Z. and Wang, J. (1983) The induction of pollen sporophytes of

winter wheat and the development of the new variety 'Jinghua N' 1'. *Sci. Agric. Sinica*, **1**, 29–35.

49. Friebe, B., Zeller, F. J. and Kunzmann, R. (1987) Transfer of the 1BL/1RS wheat–rye translocation from hexaploid bread wheat to tetraploid durum wheat. *Theor. Appl. Genet.*, **74**, 423–425.

50. Cattaneo, M. Qiao, Y. M. and Pogna, N. E. (1991) Embryoid induction and green plant regeneration from cultured anthers in a durum wheat line homozygous for the 1BL/1RS translocation. *J. Genet. Breeding*, **45**, 369–372.

51. Sanford, J. C., Klein, T. M., Wolf, E. D. and Allen, N. (1987) Delivery of substances into cells and tissues using a particle bombardment process. *Particle Sci. Technol.*, **5**, 27–37.

52. Tomes, D. T., Weissinger, A. K., Ross, M., Higgins, R., Drummond, B. J., Schaaf, S., Malone-Schoneberg, J., Staebell, M., Flynn, P., Anderson, J. and Howard, J. (1990) Trangenic tobacco plants and their progeny derived by microprojectile bombardment of tobacco leaves. *Plant Mol. Biol.*, **14**, 261–268.

53. Gordon-Kamm, W. J., Spencer, T. M., Mangano, M. L., Adams, T. R., Daines, R. J., Start, W. G., O'Brien, J. V., Chambers, S. A., Adams, Jr., W. R., Willetts, N. G., Rice, T. B., Mackey, C. J., Krueger, R. W., Kausch, A. P. and Lemaux, P. G. (1990) Transformation of maize cells and regeneration of fertile transgenic plants. *The Plant Cell*, **2**, 603–618.

54. Finer, J. J. and McMullen, M. D. (1990) Transformation of cotton (*Gossypium hirsutum* L.) via particle bombardment. *Plant Cell Rep.*, **8**, 586–589.

55. Perenzin, M., Pogna, N. E. and Borghi, B. (1992) Combining ability for breadmaking quality in wheat. *Can. J. Plant Sci.*, **72**, 743–754.

15 Quality and safety implications of genetic manipulations of food crops

R. TOWNSEND

15.1 Introduction

Biotechnology offers enormous potential to increase the quality and quantity of our food supply while decreasing chemical inputs and conserving non-renewable resources. However, the public is concerned that the technology may pose new and unfamiliar risks and looks to the regulatory agencies to examine and approve new products prior to sale. Assuring the quality and safety of food crops produced using the new technologies of genetic manipulation will be critical in promoting consumer acceptance and commercial success. Industry supports government regulation of biotechnology for public reassurance that such oversight provides and for the high standards it promotes. However, regulatory agencies, seed companies and food processors are currently struggling with the issues this raises for both newly and traditionally derived products. Assuring the safety of foods produced by genetic modification will be discussed here with reference to the recommendations of the International Food Biotechnology Council.

15.2 Safety of genetically engineered foods

The majority of the world's food, feed and oilseed crops (including maize, rice, soybean, sunflower and canola) have succumbed to the wiles of the genetic engineer. Wheat has proved to be one of the more intractable species. We have only recently seen published reports of the regeneration of fertile plants from wheat protoplasts and the establishment of stably transformed lines of wheat callus. It can only be a matter of time before these two key developments are brought together, and the first stably transformed and fertile lines of wheat are obtained.

Stable transformation of wheat will open the way to a host of genetic modifications aimed at improving pest and disease resistance, modifying processing characteristics and improving nutritional properties. Along with such developments will come questions concerning the health and environmental safety of these new varieties. Failure to provide answers that satisfy farmers, millers, food processors and consumers will prejudice the commercial potential of these products.

It seems ironic that crops developed using the highly specific and directed changes brought about by the genetic engineer are the subject of so much concern, while the safety of varieties developed by the plant breeder (who mixes entire genomes and then selects the most favoured phenotype) pass virtually unquestioned. The plant breeder enjoys a high level of public trust, because this somewhat random approach to genetic improvement has proved to be safe.

This enviable record of safety has been maintained despite the documented presence of more than 209 natural toxicants in our food crops. Twenty-one of these toxicants are known to cause adverse effects in the normal human diet. There have been a very few incidents of toxic products that were accidentally derived from plant breeding programmes. The most frequently cited example is the infamous Lenape potato cultivar, a blight resistant variety that contained elevated levels of toxic glycoalkaloids called solanines. Either plant breeders are incredibly fortunate to have avoided a major poisoning incident or there is very little genetic potential within our food crops to develop varieties expressing toxins at a level sufficient to be harmful to humans. In view of the millions of crosses made by plant breeders and the thousands of new varieties developed, the latter explanation seems to be more likely. Our traditionally derived crop varieties are therefore the standard against which the new genetically engineered products should be judged.

Other genes for insect control may come from more familiar sources. Cowpea trypsin inhibitor slows the development of lepidopteran larvae that feed on transgenic tobacco plants and Pioneer has recently shown that certain plant lectins show significant growth inhibiting activity, *in vitro*, against a major pest of maize, the European corn borer. It is too early to tell how effective these proteins will be in controlling insect pests in the field. However, genes that are isolated from traditional food crops, having a long history of safe human consumption, should raise relatively few concerns about possible toxicity.

Resistance to many plant viruses can be engineered into plants by introducing a copy of the viral coat protein gene. Resistance to barley yellow dwarf virus will be one of the first targets for biotechnologists working with wheat. Since the virus commonly infects large acreages of the crop, humans and animals must have consumed large quantities of the virus and, therefore, the viral coat protein, without experiencing any ill effects. Animal feeding studies to establish the safety of viral coat proteins would seem to be unnecessary.

In the future, most applications of genetic engineering in crops are likely to be intraspecies gene transfers. The science of genetic mapping stands poised to revolutionize the process of cultivar improvement. Plant breeders will map portions of the chromosome that encode agronomically important traits, such as drought tolerance, using restriction fragment length polymorphisms (RFLPs). Then they will use genetic engineering to move those blocks of genes into varieties deficient in those traits. These intraspecies gene transfers are equivalent to traditional plant breeding, but are much faster and more controlled and directed.

Transfers between plants of the same species or those that are sexually compatible pose no greater risk than the products of hand pollinations. Therefore, we should not make the process of genetic engineering the focus for supervision.

Critics contest this principle of 'product not process'. They suggest the process of genetic engineering is what the public is concerned about and should be the focus of regulation. The experience with irradiated foods shows that once stigmatized by process-based regulation and associated labelling requirements, the products of biotechnology are unlikely to find favour with the consumer. It is a sad reflection on our level of science education that technological developments leading to a safer more nutritious food supply are viewed with such suspicion.

Proponents of process-based regulation also point to the potential for unrelated changes or pleiotropic effects to occur as a result of genetic engineering. These include the disruption of gene function through insertion of foreign DNA, recombination between foreign DNA sequences and increased endogenous gene expression in the proximity of a strong foreign gene promoter. Some of these events have been documented as a result of *Agrobacterium* Ti-plasmid insertions where they have resulted in altered phenotypes. The concern is that such events could eliminate essential nutrients, produce elevated levels of toxic products or even the *de novo* synthesis of new toxins.

Certainly, the potential for unwanted pleiotropic effects to occur cannot be discounted, but we should view the potential risk in the context of our experience with traditional products. The plant genome is subject to a variety of genetic phenomena including chromosome breakage and unequal crossing over during cell division, gene amplifications in response to environmental stress and deletions, and amplifications and translocations associated with the activation of transposable elements. All these phenomena can lead to altered phenotypes. Indeed, plant breeders have exploited such variation in their breeding programmes and have developed methodology to increase the frequency of altered phenotypes, such as exposure to ionizing radiation and the use of tissue culture to derive somaclonal variants. Many of the resulting varieties have found their way into the food supply without any adverse consequences.

15.3 Transposable elements

Of the natural mechanisms capable of inducing genetic modifications, the insertion of transposable elements is most analogous to the introduction of foreign DNA into the plant genome. Transposons are mobile genetic elements, the so called 'jumping genes' first identified by Barbara McClintock in maize. These elements have subsequently been identified in microbes and animals, as well as a number of other plant species. Transposable elements typically contain at least one gene together with a strong promoter and terminator. In this respect, they resemble many gene vectors synthesized by genetic engineers.

At least nine different transposable elements have been identified in maize. A single plant can contain as many as 50 copies of a single transposon, although not all these will be active at one time. Transposons can be found in some of the progenitors of our most successful hybrids. Despite the presence of these elements and their potential to disrupt gene function, the breeding of hybrid maize has proceeded without incident for 70 years. There seems to be very little risk that genetic engineering will induce unique pleiotropic effects in maize which have not already occurred as a result of transposition.

15.4   US government regulations

The principle that the products of biotechnology do not pose an inherently greater risk than their traditional counterparts received official endorsement with the publication of the Office of Science and Technology Policy's (OSTP) Coordinated Framework for the Regulation of Biotechnology (*Federal Register*: Thursday, June 26, 1986). The OSTP concluded that existing statutes were adequate to address the regulatory needs of biotechnology products.

Unfortunately, many of the details concerning agency supervision of different products remain to be resolved. Most environmental releases of experimental plant varieties have been regulated by the US Department of Agriculture (USDA) under the *Federal Plant Pest Act*. Applications to field test genetically engineered plants expressing pesticidal genes are reviewed jointly by USDA and the Environmental Protection Agency (EPA). The EPA has indicated that it might regulate the commercialization of such plants under the *Federal Insecticide, Fungicide and Rodenticide Act* (FIFRA). This implies that tolerances would have to be set for the gene products under the provisions of the *Federal Food, Drug and Cosmetics Act* (FFDCA).

Such supervision may be appropriate in the case of products such as *Bacillus thuringiensis* microbial insecticide (Bt) (which is already regulated under FIFRA) or 'novel active ingredients' like scorpion venom. It is more difficult to see why an insecticidal gene isolated from one line of maize and introduced into another line of maize should be regulated as a pesticide when the traditionally derived insect tolerant lines are exempted from regulation. The necessity to set tolerances for gene products that are already part of our food supply also seems questionable.

Existing laws that ensure the wholesomeness of our food supply place the burden of proof of safety on the manufacturer. The Food and Drug Administration (FDA) has premarket clearance authority over food additives under the FFDCA. Products that are foods and not food additives can be marketed without any premarketing clearance from the Agency. New crop varieties have traditionally been viewed as Generally Regarded as Safe (GRAS) and are not subject to premarket notification or review. The incident with the Lenape potatoes caused the FDA to consider if changes in food composition that might result from traditional genetic modifications of crop plants could affect the GRAS status of the resulting foods.

The Agency indicated that an increase in a toxic moiety of more than 10% compared to the parent containing the least toxicant, or a decrease in an essential nutrient of more than 20%, could require submission of a formal GRAS affirmation petition for the new variety. The vast majority of new plant varieties have not required GRAS affirmation or any premarket review by the Agency. Plant breeders and processors, who are familiar with potential toxicants, routinely screen new varieties for the content of toxins such as glucosinolates in oil-seed rape, solanines in potatoes and tomatoes, and trypsin inhibitors in soybeans. Breeding material with unacceptably high levels of such compounds is discarded.

15.5   IFBC report

In 1988, a group of 30 or so companies, including Pioneer, representing biotechnology, plant breeding and food processing interests came together under the auspices of the International Food Biotechnology Council (IFBC). The aim of the IFBC was to develop criteria and procedures that could be used to evaluate the safety of foods produced through genetic modification. The committee brought together a group of experts from academia and industry to write a comprehensive science-based treatise,[1] aimed at regulatory agencies, the biotechnology industy, the food industry, the general public and officials at all levels of government. The report (published in 1990) has been very well received, both in the US and abroad.

The IFBC first looked at traditional foods and traditional methods of genetic modification, since these define our current standard of food safety. The report found that the screening and testing that are part of the traditional development of new plant varieties should not be relaxed, but no new regulatory measures are needed for food and food ingredients derived from plants that are genetically modified using traditional procedures. The report went on to show how an understanding of genetic modification and the results of those modifications allow the safety and regulatory issues associated with new food and food ingredients to be placed in perspective.

Particularly important are natural toxicants that may occasionally be significant sources of human hazard and should be the first priority of safety evaluation. The report acknowledges that while recombinant DNA methodology using known genes and vectors provides greater confidence than traditional methods of genetic modification, the newer techniques also retain a potential for undirected and undesired genetic and compositional changes.

The report makes certain general conclusions and recommendations for the safety evaluation of foods produced using non-traditional techniques of genetic modification. A flexible tiered approach to safety evaluation is proposed that is guided by decision trees. The key data required to make decisions rely on a mix of genetic, compositional and toxicological information. The report concludes that it will rarely, if ever, be necessary to pursue any of these three aspects to exhaustion. 'We cannot regulate everything that we should be concerned about, and we cannot be concerned about everything.' There must be a threshold for regulation or even concern below which further evaluation need not be conducted.

The IFBC procedures for safety evaluation are embodied in a series of decision trees: one deals with ingredients; one with single chemicals and simple mixtures; and one with complex mixtures and whole foods. It is this last decision tree that most concerns those who develop new plant varieties and utilize plant products to make processed foods. Comparison of the new product with its traditional counterpart remains the premise of the evaluation process. Particular attention should be focused on nutrient content, other desired or unexpected expression products and any toxic constituents of the host organism. Beyond this, it is reasonable to require documentation of the genetic change, a determination of exposure and possibly some appraisal of palatability.

The decision tree asks a number of key questions involving the source of genetic material. The use of genetic material from traditional food should almost always provide greater confidence in the safety of the product than the use of genetic material with which we have no dietary experience. It is also anticipated that there will be an expanding list of acceptable genetic elements as new products are developed.

The decision tree next addresses the composition of the food. Compositional screening should normally be limited to any constituents intentionally introduced or modified, any constituents of nutritional or safety significance likely to vary in concentration as a result of the modification and other inherent constituents.

Inherent constituents are those components that are essential nutrients or naturally occurring toxins that are normally present in that species. The standard for compositional comparison must be the range that is normal in any closely related traditional food or food product.

Finally, the decision tree turns to toxicological data. When the new food contains sufficient quantities of some constituent or constituents with no dietary history of safe use, then some toxicological testing may be necessary to ensure safety. In some instances, it may not be possible to isolate sufficient quantities of a gene product to conduct feeding studies with the purified substance. In those cases, feeding of the whole food may be appropriate. Feeding studies would normally be of short duration to screen for acute or subchronic effects to detect the presence of unexpected toxicants that have escaped detection by other means.

The report concludes with a detailed review of the current legal and regulatory framework for ensuring the safety of food in the United States. The IFBC recommends that regulation of genetically modified foods should be patterned directly on existing law and practice. The report further recommends that the FDA consider adopting some form of flexible voluntary procedures for informing the agency about the applications of biotechnology that might not require formal FDA review. This would help to keep the agency informed about new technologies and products and contribute to public and market confidence in those products.

In May 1992, FDA issued a policy entitled *Foods Derived from New Plant Varieties (Federal Register* **57**, 22984–23005). The Agency stated that it would regulate genetically modified plants within FDA's existing framework and that the regulatory status of the food, irrespective of the method by which it was developed, is dependent upon the objective characteristics of the food and the intended use of the food. Pre-market review and approval by the Agency would only be required if there were questions of safety sufficient to bring into question the GRAS status of the food. FDA also provided guidance for developers in reviewing the scientific considerations for evaluating the safety of their products based on existing practices followed by traditional plant breeders. The Agency stated that it was not the intention to alter these long established practices, or to create new regulatory obligations for them. The FDA provided a series of decision trees to aid developers in reviewing their products and deciding if and when consultation with the Agency was appropriate.

Reference

1. International Food Biotechnology Council. (1990) Biotechnologies and food: assuring the safety of foods produced by genetic modification. *Regulatory Toxicol. Pharmacol.*, **12**(3).

Index